Digital Technologies for Engineering Construction Information Productivity

Digital Technologies for Engineering Construction Information Productivity goes beyond other digital construction books by examining how emerging digital construction technologies and other popular technologies developed by, for example, Microsoft can be used in improving the productivity of construction information.

The book discusses both the theory and practice through use-case illustrations in integrated software environments and successfully blends applications and workflows of construction technologies with more common and non-industry-specific software systems. It seeks to unleash the power of readily available digital technologies in optimizing the productivity of construction project information management.

This book is essential for students, academics, educators, construction professionals, and construction software developers. For students, the book provides a solid foundation in understanding digital construction technologies including building information modelling (BIM) and some selected Microsoft Products and their applications in managing construction project information. For academics and educators, it will provide insights into research niches around integration of the different digital technologies and serve as a useful text for teaching at undergraduate and postgraduate levels in Architecture, Engineering, and Construction (AEC) subjects. Professionals will find this useful for the adoption of digital technologies in their own practice while software developers will find this beneficial in the development of interoperability software for efficient information exchange between the various software systems.

Fonbeyin Henry Abanda is an Associate Professor at the *École Nationale Supérieure des Travaux Publics* in Yaoundé, Cameroon, and formerly served as a Reader and Associate Professor in Construction IT at the School of the Built Environment, Oxford Brookes University, United Kingdom. His research interests are in the areas of Semantic Web, Digital Construction Technology, and Artificial Intelligence. He has worked on research projects funded by the Engineering and Physical Sciences Research Council, the International Labour Organisation and the Intergovernmental Panel on Climate Change. He has designed, implemented and delivered BIM related modules on the undergraduate and postgraduate courses in the School of the Built Environment, Oxford Brookes University. He is currently supervising a number of PhD students working on construction project management, facilities management, BIM, Digital Twins and the Semantic Web. Henry has delivered BIM lectures in so many countries including India, Ecuador, Cameroon, Kenya, Ethiopia, Ghana, Nigeria, Spain, France, UK, Canada, Finland, Germany, The Netherlands, etc. Henry was the lead facilitator in training 180 construction professionals through hands-on workshops on the use of BIM for construction project management and quantity surveying in the UK.

Digital Technologies for Engineering Construction Information Productivity

Fonbeyin Henry Abanda

Routledge
Taylor & Francis Group
LONDON AND NEW YORK

Designed cover image: Fonbeyin Henry Abanda

First published 2026
by Routledge
4 Park Square, Milton Park, Abingdon, Oxon OX14 4RN

and by Routledge
605 Third Avenue, New York, NY 10158

Routledge is an imprint of the Taylor & Francis Group, an informa business

British Library Cataloguing-in-Publication Data
A catalogue record for this book is available from the British Library

ISBN: 978-1-032-76038-4 (hbk)
ISBN: 978-1-032-75874-9 (pbk)
ISBN: 978-1-003-47674-0 (ebk)

DOI: 10.1201/9781003476740

Typeset in Times New Roman
by KnowledgeWorks Global Ltd.

To the cherished memory of my beloved parents, Woghombong Abanda and Nyieba Martina. Your unwavering love, timeless wisdom, and countless sacrifices continue to guide and inspire every step of my journey. Though you are no longer with me in body, your spirit lives on in all that I strive to become.

With deepest affection and pride, I dedicate this work to my precious children: Farnyi Louibeuka Abanda, Mason Wofarnyi Abanda, and Adriel Abenyi Abanda. You are the heartbeat of my aspirations, the embodiment of my legacy, and the reason I press forward with purpose. May this work serve as a beacon, reminding you of the boundless possibilities unlocked by knowledge, perseverance, and faith.

It is my earnest prayer that each of you will go on to achieve abundantly – 70 × 7 what I have accomplished – and rise to distinction as Professors and thought leaders in your respective fields. May your lives be marked by excellence, integrity, and an enduring commitment to uplift others.

Contents

About the Author

Fonbeyin Henry Abanda, HDR, PhD, Dipl.-Ing., PGcert (Distinction), BSc (Hons), CEng, MIET, FHEA, is Maître de Conférences in the Ecole Nationale Supérieure des Travaux Publics, Yaoundé-Cameroon and formerly a Reader/Associate Professor in Construction IT at the School of the Built Environment, Oxford Brookes University, UK. With an extensive academic background that spans multiple disciplines and institutions, Dr. Abanda holds a BSc (Hons) in Mathematics and Physics from the University of Buea, Cameroon; a Dipl. Ing. in Civil Engineering from École Nationale Supérieure Polytechnique de Yaoundé, Cameroon; an MA in Monitoring & Evaluation from Africa Nazarene University, Kenya; a PGCert (Distinction) in Teaching in Higher Education from Oxford Brookes University; and a PhD in Construction IT from the same institution. He has also earned an HDR ('Habilitation à Diriger des Recherches') in Digital Construction Technology from Institut National Polytechnique de Toulouse, France.

Dr. Abanda's research and teaching focus on advanced areas such as the Semantic Web, Digital Construction Technology, and Artificial Intelligence. His work has garnered recognition and funding from prestigious organizations, including the Engineering and Physical Sciences Research Council, the International Labour Organization, and the Intergovernmental Panel on Climate Change.

At Oxford Brookes, Dr. Abanda has been instrumental in designing and delivering BIM-related courses for both undergraduate and postgraduate students within the School of the Built Environment. He currently supervises multiple PhD students working on cutting-edge topics like construction project management, facilities management, BIM, Digital Twins, and the Semantic Web.

A true BIM enthusiast, Dr. Abanda has emerged as a leading voice in promoting BIM adoption worldwide. He has spearheaded a massive campaign to raise awareness about BIM, engaging audiences both remotely and face-to-face in numerous countries. His efforts have made him a recognized BIM evangelist, passionately advocating for the integration of digital technologies in the construction industry.

Dr. Abanda is also a global thought leader, having delivered lectures and workshops across five continents, including countries such as India, Ecuador, Nigeria, Spain, France, Cameroon, Kenya, Ghana, and the UK. As a lead facilitator, he has trained over 180 construction professionals in the UK through hands-on workshops on BIM for construction project management and quantity surveying.

Beyond his academic and research contributions, Dr. Abanda serves as one of the five Strategic Advisors for BIM Africa, a pivotal initiative driving the adoption and regulation of digital technologies in the Architecture, Engineering, and Construction industries across the African continent.

Preface

In today's fast-paced and interconnected world, digital technologies are reshaping every facet of our lives, and the engineering and construction industry is no exception. As a vital pillar of global development, construction engineering demands efficiency, precision, and innovation. The integration of digital technologies into this dynamic sector has ushered in a new era of unprecedented possibilities, redefining the way we plan, design, construct, and manage our built environment.

This book is a comprehensive exploration of how the fusion of cutting-edge digital tools and engineering principles is revolutionizing the construction landscape. As we stand at the brink of a digital renaissance, this book serves as a guiding beacon, illuminating the path towards enhanced productivity, sustainable practices, and exceptional project outcomes.

In the ensuing chapters, we embark on an enlightening journey through the vast realm of digital technologies available to the engineering construction industry. From BIM, emerging technologies (drones, 3D printers, 3D laser scanners, etc.), and business intelligence systems, for example, Microsoft Power BI, each chapter uncovers the power of these transformative tools and their potential impact on information productivity.

Currently, most digital construction technology books are too theoretical with less emphasis on practical applications. On the other hand, those that are practical tend to focus on a single specific software. While the details of studying specific software are important, the rapid and regular updates of the software versions easily make these books outdated. Focusing on any one software fails to illuminate a bigger picture on the flow of information between systems, people, or software. Thus, the workflows depicting flow of information between systems and actors have hardly been discussed in most books. Lastly, many digital technology concepts, for example, benefits of BIM, are often presented in haphazard manner without aligning them to standard construction project lifecycle.

Our goal is to overcome the aforementioned weaknesses and equip readers with the knowledge and tools needed to harness the full potential of digital construction technologies for managing construction information. With insightful practical examples, step-by-step guides, workflows, and adoption of a system approach using RIBA plan of work 2020 project lifecycle, this book caters to professionals at every level of the construction engineering spectrum. Whether you are an architect, civil engineer, project manager, or technology enthusiast, this book will empower you to embrace digital transformation and maximize your productivity in managing construction information.

As we journey through the pages of this book, we emphasize the need for collaboration and cross-disciplinary integration. The power of digital technologies lies not in isolated applications, but in their seamless convergence, empowering us to unlock new horizons of innovation and efficiency.

We recognize that embracing digital transformation is not without its challenges. The construction industry's unique complexities and traditions may initially pose hurdles to digital adoption. However, our unwavering belief is that with the right knowledge and the right mindset, every obstacle can be overcome.

As the construction engineering industry embarks on a transformative digital journey, we invite you to join us in this pursuit of excellence. Let us harness the power of digital technologies to optimize construction processes, minimize waste, and create a built environment that aligns with our vision of a sustainable and prosperous future.

Acknowledgements

I wish to express my deepest and most heartfelt gratitude to Professor Pettang Chrispin, my supervisor, lecturer, and mentor at *École Nationale Supérieure Polytechnique (ENSP), Yaoundé*. Your steadfast guidance, wise counsel, and generous support have been pivotal to my academic and professional growth. Thank you for being a beacon of excellence and integrity throughout my journey.

A special and sincere appreciation goes to Professor Marcelline Blanche Manjia, my former classmate and now the esteemed Head of the Department of Civil Engineering at ENSP Yaoundé. Your tireless commitment to education, your exemplary leadership, and your personal encouragement continue to be a source of profound inspiration.

My heartfelt thanks also go to Professor George Elambo Nkeng, Director of *École Nationale Supérieure des Travaux Publics (ENSTP), Yaoundé*, whose visionary leadership and dedication to nurturing academic excellence have left a lasting impression on my path.

I would also like to lovingly acknowledge my dear office colleagues at ENSTP Yaoundé – Nformi Beatrice Maluh, Egbenchong Ruth Eneke, Etengeneng Johannes Ojong, and Djoukwo Tsanetse Majolie Carine. Your camaraderie, insightful exchanges, and the warmth of our shared space added light and laughter to my days. Thank you for making our office not just a place of work but a haven of support and motivation.

I am profoundly grateful to Fidèle Kengne, the visionary founder of *Rousseau Higher Institute of Technology, Douala*. Your unwavering belief in my abilities and your constant encouragement ignited the spark that gave birth to this book. Your faith in the power of education and innovation continues to inspire my efforts.

To the incredible students of Rousseau Higher Institute of Technology, Douala, *you are my daily inspiration and greatest source of joy*. Witnessing the evolution of our first and second cohorts in this young institution has been a truly rewarding experience. Your rich diversity, drawn from both Francophone and Anglo-Saxon traditions, has profoundly shaped the direction and purpose of this work. This book is, in many ways, a reflection of your resilience, curiosity, and limitless potential.

Finally, to all who have contributed, whether in great measure or small, in words or in deeds, please know that your kindness, support, and encouragement have made this endeavour not only possible but also truly meaningful.

This book benefited from the use of AI tools, specifically ChatGPT-4o by OpenAI, which assisted in the drafting and content development of Sections 8.3 and 9.2. The tool was also employed to support the proofreading of the entire manuscript. All content has been carefully reviewed, verified, and finalized by the author, who assumes full responsibility for the accuracy, integrity, and originality of the work.

Chapter 1

General Introduction

1.1 Introduction

The built environment has experienced a significant rise in the adoption of digital construction technologies aimed at enhancing the performance and efficiency of construction projects, which have long suffered from high inefficiencies and low productivity. Despite this growing interest, there is a noticeable gap in the literature concerning the tracking and comprehensive analysis of these emerging digital technologies and their practical applications. Many of these technologies are still utilized in isolation, without a clear understanding of how they can be integrated to maximize their potential. There is a distinct lack of books that explore a wide array of digital construction technologies, particularly those that illustrate the connections between different technologies and their integrated use in real-world projects. This proposed book aims to fill that gap by offering an extensive examination of various digital construction technologies, including building information modelling (BIM), drones, robots, 3D laser scanners, 3D printers, and cloud-based BIM, among others, along with their integrated applications in practice. The book is excellent and will go a long way to meet the needs of readers.

This chapter sets the stage for a comprehensive exploration of how digital construction technologies can revolutionize the construction industry, particularly in addressing the critical issues of housing productivity and affordability. It begins by outlining the book's core philosophy, which emphasizes the integration of digital tools to overcome inefficiencies. It then provides a global perspective on the construction industry's current state, followed by an analysis of housing productivity and affordability challenges. The chapter continues by discussing various strategies for providing affordable housing, such as sustainable materials and renewable energy, before critically examining the obstacles that hinder these efforts. Finally, it positions digital construction technologies as a pivotal solution to these challenges, highlighting their potential to enhance efficiency, reduce costs, and ultimately deliver more sustainable and accessible housing worldwide.

1.2 The Underpinning Philosophy of the Book

In parallel with the rapid introduction and proliferation of various digital technologies in the market, the language within the industry has also evolved at an accelerated pace. Recognizing this shift, the chapters in this book have been carefully crafted to reflect the current terminology and industry practices. This ensures that the language used is aligned with contemporary industry standards and thinking. Specifically, when discussing BIM standards, particular emphasis has been placed on the terminologies outlined in the most recent ISO 19650-1 and ISO 19650-2

DOI: 10.1201/9781003476740-1

standards. Additionally, we have highlighted the transition and changes from the older PAS 1192-2:2013 standard to these newer ISO standards. This approach is intended to help readers grasp the evolution of digital construction technologies over the years, providing them with a clear understanding of what is currently relevant and required in the industry.

Moreover, the importance of historical knowledge in predicting future trends is underscored by the fact that many practitioners continue to use outdated terminologies and concepts from the PAS 1192-2:2013 standard, which are still prevalent in the open-source literature. By tracing the evolution of these terms and concepts, readers can better appreciate the progression of the field and understand the rationale behind the adoption of new standards.

Given the process-driven nature of construction activities, process maps and workflows have been incorporated to facilitate the implementation of digital construction technologies. These visual tools simplify the understanding and application of these technologies in practice. Additionally, to enhance comprehension and ensure a thorough evaluation of digital construction technologies, we have aligned certain aspects, such as the benefits of these technologies, with the UK RIBA Plan of Work 2020 project lifecycle standard. This alignment provides a structured framework that bridges the gap between theoretical knowledge and practical application, ensuring that readers can effectively integrate digital construction technologies into their projects.

1.3 Global Perspectives of the Construction Industry

The construction industry is often regarded as a barometer of a country's progress, reflecting not only economic vitality but also the ability to meet the basic needs of its population. The industry's contribution to a nation's gross domestic product (GDP) is significant, but its role extends beyond mere economic metrics – it is also a critical determinant of social stability and quality of life. However, one of the most pressing global challenges facing the industry today is the provision of adequate housing and infrastructure to keep pace with the explosive population growth and rapid urbanization occurring worldwide. By 2050, the global population is projected to increase by 2.5 billion people, with urban residents comprising 68% of the total population, a substantial rise from current levels (United Nations, 2018).

This demographic shift is particularly pronounced in Africa, which is experiencing the most rapid population growth and urbanization rates in the world. In 2014, more than 200 million people in Africa lived in slums, accounting for 47% of the urban population (UN-Habitat, 2016). These statistics paint a bleak picture of the housing crisis, which is poised to become even more severe in the coming decades.

The implications of this crisis are profound, particularly when considering housing productivity and affordability. The construction industry is already under significant strain to meet current demands, and the projected population increases will only exacerbate these pressures. The industry's productivity has historically lagged behind other sectors, largely due to its reliance on traditional building methods, insufficient investment in innovation, and a fragmented approach to project management. As the demand for housing skyrockets, the limitations of these outdated practices become increasingly apparent.

Moreover, the challenge of affordability further complicates the situation. In many parts of the world, especially in developing countries, there is a stark disconnect between the cost of constructing new homes and what is affordable for the average citizen. This affordability gap is widening as construction costs rise due to factors such as increasing material prices, labour shortages, and regulatory constraints. As a result, even when new housing units are built, they often remain out of reach for those who need them most.

The failure to address these issues will have far-reaching consequences. A continued shortage of affordable housing will likely lead to the expansion of informal settlements and slums, perpetuating cycles of poverty and social inequality. Furthermore, inadequate housing conditions are closely linked to a host of other problems, including poor health outcomes, reduced educational attainment, and increased vulnerability to environmental hazards. The construction industry's inability to keep up with housing demand could, therefore, exacerbate social unrest and economic instability in regions already under pressure.

To address these challenges effectively, the construction industry must undergo a fundamental transformation, prioritizing greater efficiency, innovation, and sustainability. However, before we can delve into the specific shifts needed to enhance housing productivity and affordability, it is essential to first develop a clear understanding of these critical concepts.

1.4 Housing Productivity and Affordability

Housing is universally recognized as a fundamental human right, essential for ensuring personal security, privacy, and health. Despite this, the global challenge of ensuring adequate housing productivity and affordability remains a significant barrier, preventing many from accessing decent living conditions. Housing productivity is often defined as the efficient use of resources, processes, and technologies in the construction and maintenance of housing units. However, this definition can be critiqued for its narrow focus on efficiency and cost minimization, often overlooking the critical aspect of quantity. True housing productivity should not only encompass the optimization of construction methods and materials but also the ability to produce a sufficient number of housing units to meet the growing demand, all while ensuring sustainability and long-term resilience.

The issue of housing shortages must be critically appraised from both demand and supply perspectives. On the demand side, rapid urbanization, population growth, and increasing migration have led to an unprecedented surge in the need for housing. However, the supply side has struggled to keep pace, largely due to bottlenecks in the construction industry, such as outdated building methods, regulatory hurdles, and a lack of skilled labour. This imbalance between demand and supply exacerbates housing shortages, leading to overcrowding, the proliferation of informal settlements, and a decline in living standards.

Housing affordability is another critical aspect that requires careful examination. While it is generally defined by the ability of individuals or households to obtain housing without compromising their ability to meet other essential living expenses, this definition varies widely across different contexts. The commonly cited benchmark that affordable housing should cost less than 30% of a household's income (Riverview, 2024) is a useful guideline but fails to capture the complexity of affordability in different economic environments. Factors such as location, access to services, and the overall cost of living must also be considered.

There is an urgent need to increase housing productivity while ensuring that housing remains affordable. This requires a multifaceted approach that not only embraces innovative construction technologies but also addresses the underlying economic, social, and regulatory challenges that hinder progress. By enhancing productivity in the housing sector, while simultaneously ensuring that these innovations lead to cost reductions, the industry can better meet the global demand for affordable housing. Ultimately, achieving this balance is crucial for fulfilling the fundamental human right to adequate housing for all.

1.5 Strategies for Providing Affordable Housing

1.5.1 *Zero-Energy and Carbon Buildings*

With an increasing number of climate-related catastrophes globally, an urgent need to ensure the sustainability of assets including buildings is imperative. Many parameters need to be considered in the various phases of an asset's lifecycle. Some of the parameters or factors, e.g., operational energy, embodied energy, and greenhouse gas emissions, need to be thoughtfully considered during each of the asset's lifecycle phases. In order to ensure the sustainability of assets, the various performance parameters should be optimized. Optimization which minimizing or maximizing the parameters is often referred to as efficiency. With respect to buildings and energy, the goal of minimizing or reducing the amount of energy needed to service or run a building is known as building energy efficiency. Many different concepts have been used to promote the efficiency of buildings. The first is net-zero energy buildings, which refers to a building with net energy consumption equal to zero. From an emission perspective, a similar concept exists often referred to as net-zero carbon-zero. Its equivalents are carbon-zero and carbon-neutral buildings.

Different organizations and countries have developed standards and encouraged their use in achieving net-zero energy or carbon benchmark in buildings. Some of the standards include:

- Leadership in Energy and Environmental Design (LEED)
- Building Research Establishment Environmental Assessment Method (BREEAM)
- Passive House
- International Living Future Institute (ILFI)
- China Green Building Council (CGBC)
- Australian Building Codes and the National Australian Built Environment Rating System (NABERS)
- Singapore Green Building Council (SGBC)
- European Union (EU) Directives
- Indian Green Building Council (IGBC)

Adopting a net-zero energy initiative which can be enforced by using the aforementioned standards can contribute to more affordable housing in several ways:

- **Lower Maintenance Costs**
  - **Durability of Materials:** Net-zero buildings often incorporate high-quality, durable materials and advanced construction techniques. This results in a building that is not only more energy-efficient but also requires less maintenance over time. Reduced maintenance translates directly into cost savings for homeowners and landlords, which can be passed on to residents in the form of lower rents or lower maintenance fees.
  - **Efficient Systems:** The use of highly efficient HVAC systems, water heating, and electrical systems means that these components are less likely to break down or require frequent repairs, further reducing the overall maintenance costs associated with the building.
- **Increased Resilience**
  - **Climate Adaptation:** Net-zero buildings are designed with climate resilience in mind, meaning they are better equipped to handle extreme weather events, from heatwaves to

cold snaps, without significant damage. This resilience can prevent costly repairs and insurance claims, ultimately contributing to lower housing costs.

- **Energy Independence:** By generating their own energy, net-zero homes are less susceptible to power outages and can avoid the cost spikes associated with grid-supplied electricity. This independence not only makes homes more resilient but also ensures stable, predictable utility costs.

- **Long-term Savings**
  - **Return on Investment:** The initial cost of integrating net-zero features, such as solar panels and high-performance insulation, is often offset by the long-term energy savings. Over time, the reduction in energy bills can result in significant savings, making housing more affordable in the long run.
  - **Increased Property Value:** Properties with net-zero capabilities often have higher resale values. While this might not directly translate to affordability, it does mean that the investment in net-zero technologies is likely to pay off, making it a financially sound decision that supports the broader goal of affordability through reduced lifetime costs.
- **Reduced Operating Costs**
  - **Lower Energy Bills:** The most direct impact of net-zero buildings on affordability is the drastic reduction or complete elimination of energy bills. By producing as much energy as they consume, net-zero homes can significantly lower the cost of living for residents.
  - **Water Efficiency:** Many net-zero initiatives also focus on reducing water consumption through efficient fixtures and landscaping. Lower water usage not only conserves an essential resource but also reduces utility bills, further lowering the cost of living.

Digital construction technologies can play a significant role in ensuring that the aforementioned standards can be implemented in affordable housing development strategies. The strategies for providing affordable and sustainable housing and infrastructure to meet the ever-rising demand from a global growing population will be examined in the ensuing section.

1.5.2 Construction Using Sustainable Construction Materials

Considering that building materials constitute approximately 80% of the total cost of construction (Kayode and Olusegun, 2013), the choice of materials is a critical determinant of housing affordability. In many regions around the world, there has been a heavy reliance on conventional materials and products that are widely used in developed countries. These materials, often imported, carry high costs due to transportation, import taxes, and the need for specialized labour, which in turn drives up construction expenses and reduces affordability. Additionally, these materials may not be well-suited to the environmental conditions in different parts of the world, leading to inefficiencies, higher maintenance costs, and sustainability challenges.

The use of locally sourced, sustainable construction materials offers a significant opportunity to improve housing affordability globally. Materials such as earth, bamboo, straw, and other naturally occurring resources can be obtained locally at a fraction of the cost of imported materials. These materials generally require lower levels of energy for extraction, processing, and transportation, which further reduces their overall cost. Moreover, the use of local manpower and technology in the production and application of these materials helps to develop local expertise, create jobs, and stimulate regional economies, all of which contribute to reducing overall construction costs.

Globally, innovative local materials are gaining recognition for their affordability and sustainability. In Latin America, for example, adobe and rammed earth construction techniques have been used for centuries and are experiencing a resurgence due to their low cost and excellent thermal properties. In Mexico, communities have successfully built affordable homes using these materials, which provide natural insulation against both heat and cold, reducing the need for energy-intensive heating and cooling systems. This not only makes the homes more affordable but also aligns with sustainability goals by lowering energy consumption.

In Asia, bamboo is widely used as a construction material, particularly in countries like India and Indonesia. Bamboo is not only abundant and renewable but also highly durable and flexible, making it an ideal material for affordable housing in earthquake-prone areas. For instance, in Bali, Indonesia, bamboo has been utilized in constructing affordable, eco-friendly homes that are both aesthetically pleasing and resilient to natural disasters. The use of bamboo in these regions has helped to reduce construction costs significantly, while also promoting sustainable building practices.

In Europe, particularly in countries like Spain and Portugal, the use of straw bale construction is gaining traction as a cost-effective and sustainable alternative to conventional building materials. Straw bales, which are agricultural by-products, are inexpensive, provide excellent insulation, and have a low environmental impact. Projects in these countries have demonstrated that straw bale homes can be built at a fraction of the cost of traditional homes, without compromising on quality or durability. These homes also offer significant energy savings over time, further enhancing their affordability.

The growing recognition and use of sustainable materials across different continents are also leading to the development of more resilient and environmentally friendly housing. For example, in North America, the use of reclaimed wood and recycled steel in construction is becoming increasingly popular. These materials not only reduce waste and the demand for new raw materials but also lower construction costs, making housing more affordable. In Canada, for instance, the use of reclaimed wood in affordable housing projects has helped to preserve natural resources and reduce construction expenses, while also adding unique character to the homes.

The global shift towards using sustainable, locally sourced materials also reduces dependency on volatile global markets for construction inputs. By developing and scaling the use of local materials, countries can mitigate the risks associated with price fluctuations and supply chain disruptions that often affect imported materials. This stability in material costs is crucial for maintaining affordability over the long term.

Moreover, the environmental benefits of using sustainable materials are universal. Reduced reliance on high-energy materials, such as cement and steel, and the promotion of renewable resources contribute to lower carbon emissions and more sustainable building practices. This alignment with global sustainability goals can attract international funding and support for projects that prioritize affordability and environmental responsibility.

In summary, the use of sustainable construction materials is a key factor in enhancing housing affordability worldwide. By leveraging local resources, reducing dependency on costly imports, and fostering innovation in material use, the construction industry can make significant strides towards providing affordable, sustainable housing solutions. This approach not only addresses the immediate needs of housing affordability but also contributes to broader economic development and environmental sustainability across continents.

1.5.3 *Adoption of Renewable Energy*

In the face of climate change, many governments and international organizations have implemented policies and strategies aimed at mitigating environmental impacts, particularly in developing countries. A key component of these strategies is the promotion of renewable energy technologies for generating clean energy. Initiatives like the Clean Development Mechanism (CDM) are designed to support clean energy projects in developing countries, helping to reduce greenhouse gas emissions. Renewable energy is energy derived from sources that are naturally replenished or continuously available, unlike finite fossil fuels. While there is some debate over what qualifies as renewable energy – nuclear energy being a prime example of contested sources – the consensus typically includes sunlight, wind, water, biomass, geothermal, and tidal energy as the most recognized renewable sources.

The technologies associated with these renewable energy sources – such as solar panels, wind turbines, hydroelectric systems, combined heat and power (CHP) systems, geothermal systems, and tidal energy systems – are increasingly being adopted worldwide. Solar energy, for example, harnesses power from the sun and can be utilized through passive heating, solar thermal systems, or photovoltaic panels. Wind energy systems capture energy from wind motion, while hydroelectric systems generate power from the movement of water. CHP systems can efficiently produce electricity and heat from biomass or natural gas, provided that the carbon footprint is managed through replanting efforts. Geothermal systems tap into the earth's core for energy, and tidal systems exploit the movement of tidal waters to generate electricity.

The integration of these renewable energy technologies into housing can significantly contribute to making housing more affordable, particularly for low- and moderate-income families. One of the primary ways this is achieved is through **reduced energy costs**. By utilizing renewable energy sources, homeowners can lower their reliance on traditional energy grids, which are often expensive and subject to price fluctuations. For instance, solar panels can reduce or even eliminate electricity bills for households, making homeownership more affordable over the long term. In regions where energy costs are a significant portion of household expenses, this reduction can make a substantial difference in the overall affordability of housing.

Energy efficiency is another crucial factor. Homes designed or retrofitted with renewable energy technologies often include energy-efficient appliances and systems that further reduce energy consumption. For example, a house equipped with a solar thermal system for water heating will use less energy than one relying on conventional electric water heaters. The upfront investment in renewable energy systems can be offset by the long-term savings from reduced energy bills, contributing to the overall affordability of the home.

Renewable energy also enhances the **resilience and reliability** of housing. In areas prone to natural disasters or frequent power outages, renewable energy systems, such as solar panels with battery storage, can provide a reliable and continuous power supply. This not only ensures that homes remain functional during emergencies but also reduces the costs associated with power outages, such as spoiled food or lost income from missed work. The resilience provided by renewable energy can thus improve the long-term affordability and security of housing.

Moreover, the adoption of renewable energy in housing aligns with broader sustainable development goals, which emphasize the need for environmentally responsible and economically viable solutions. As urban areas expand, unplanned growth can strain resources and increase the cost of living. Integrating renewable energy into new housing developments can help manage these costs by providing stable, low-cost energy solutions that do not deplete natural resources or contribute to pollution.

To conclude, the adoption of renewable energy technologies plays a critical role in enhancing housing affordability. By reducing energy costs, improving energy efficiency, and providing greater resilience, renewable energy makes housing more accessible and sustainable for low- and moderate-income families. As the world continues to grapple with the challenges of urbanization and climate change, incorporating renewable energy into housing strategies is not just an environmental imperative but also a practical solution for improving housing affordability on a global scale.

1.5.4 Adoption of Offsite Construction

In the literature, various terms have been used interchangeably to describe offsite manufacturing, including offsite industrialization (Zhai et al., 2014), manufactured construction (Arif and Egbu, 2010), offsite fabrication, pre-assembly, prefabrication, modern methods of construction, modularization, and others (Christoforou et al., 2016; Nadim and Goulding, 2011; Pan et al., 2012; Yunus and Yang, 2012). While there are minor differences in these terms based on context and usage, the general term 'offsite manufacturing' will be used for consistency. Offsite manufacturing has the potential to play a significant role in providing affordable housing by addressing several key challenges and offering various advantages that contribute to cost-effective and efficient housing solutions.

One of the primary ways offsite construction contributes to housing affordability is through **cost efficiency**. By shifting a significant portion of the construction process to a controlled factory environment, offsite manufacturing reduces material waste, optimizes resource use, and lowers labour costs. The precision of factory production minimizes errors and rework, leading to substantial savings. These cost reductions are passed on to the homebuyer, making housing more affordable. Additionally, offsite manufacturing allows for the bulk purchasing of materials, which further drives down costs through economies of scale.

Another significant advantage is the **speed of construction**. Offsite manufacturing allows for simultaneous production and site preparation, drastically reducing the overall construction timeline. Homes can be constructed in a fraction of the time it would take using traditional methods, leading to quicker occupancy and faster return on investment. This speed is particularly beneficial in regions facing urgent housing shortages, where the rapid delivery of affordable homes is critical.

Quality control is also enhanced in offsite construction. The controlled environment of a factory setting allows for rigorous quality checks at every stage of production, ensuring consistent and high-quality output. This level of quality control is difficult to achieve on traditional construction sites, where variables such as weather and site conditions can impact the final product. The result is a higher-quality home that requires less maintenance over time, reducing long-term costs for homeowners and contributing to the overall affordability of housing.

Sustainability and **energy efficiency** are additional benefits of offsite manufacturing. The precision of factory construction reduces waste and promotes the use of sustainable materials and practices. Many offsite construction projects incorporate energy-efficient designs and technologies, such as improved insulation and renewable energy systems, which reduce the ongoing energy costs for homeowners. Lower energy bills make homes more affordable over the long term, particularly for low- and moderate-income families.

Offsite manufacturing also offers the advantage of **reduced site disruption.** Since much of the construction is completed offsite, the impact on the surrounding environment is minimized. This is particularly important in urban areas where construction noise, dust, and traffic disruptions can be significant issues. Reduced site disruption not only improves the quality of life for neighbouring residents during the construction process but can also lower costs associated with site management and cleanup, contributing further to housing affordability.

Moreover, **economies of scale** play a crucial role in making housing more affordable through offsite manufacturing. As the demand for prefabricated homes increases, factories can produce components in larger quantities, reducing the per-unit cost. This scalability allows developers to offer affordable housing at a lower price point without compromising on quality or sustainability. In addition, the modular nature of offsite construction enables customization at a lower cost, allowing for the design of homes that meet specific needs while still benefiting from mass production efficiencies.

In conclusion, the adoption of offsite construction methods offers a comprehensive solution to the challenges of housing affordability. By reducing costs, speeding up construction, ensuring quality, promoting sustainability, and minimizing site disruption, offsite manufacturing creates more affordable, high-quality housing options for a wide range of populations. As housing demand continues to grow globally, especially in urban areas, offsite construction presents a viable and scalable approach to meeting this demand while ensuring that homes remain affordable and accessible to those who need them most.

1.6 Challenges Delivering Affordable Housing

1.6.1 Challenges Achieving Zero-Energy and Carbon Buildings

The concept of zero-energy and carbon-neutral buildings is an ambitious and necessary response to the increasing number of climate-related catastrophes globally. However, achieving these standards presents several challenges, particularly when integrating them into affordable housing strategies. The primary challenge lies in the initial capital costs associated with constructing zero-energy and carbon-neutral buildings. The technologies required to minimize energy consumption and carbon emissions, such as solar panels, high-performance insulation, and energy-efficient HVAC systems, often come with significant upfront expenses. These costs can be prohibitive for developers focused on affordable housing, where budget constraints are typically tighter.

Another challenge is the technical expertise required to design and build these advanced structures. Zero-energy buildings demand a high level of precision in design and construction to ensure that all systems work together efficiently. This requires skilled professionals, including architects, engineers, and builders, who are familiar with the latest sustainable building practices and technologies. In many regions, particularly in developing countries, there is a shortage of such expertise, making it difficult to implement these standards widely.

Moreover, regulatory frameworks in many countries do not yet fully support or incentivize the development of zero-energy or carbon-neutral buildings. While some countries have adopted green building standards such as LEED or BREEAM, others lack the necessary regulations to mandate or encourage the adoption of these practices. Without supportive policies, including subsidies, tax incentives, or mandates for renewable energy integration, the widespread implementation of zero-energy buildings in affordable housing remains challenging.

Finally, maintenance and long-term performance are critical concerns. While zero-energy buildings are designed to be energy-efficient and sustainable, maintaining these standards over the building's lifecycle requires ongoing management and potentially higher maintenance costs. This is particularly challenging for affordable housing, where residents may lack the resources to maintain high-tech systems, and where housing providers may struggle with limited budgets for ongoing upkeep.

1.6.2 Challenges of Hindering the Use of Sustainable Construction Materials

While the use of sustainable construction materials offers significant potential to improve housing affordability, several challenges impede their widespread adoption. One major challenge is the availability and consistency of supply. Locally sourced materials like bamboo, straw, or earth may not be available in sufficient quantities or with consistent quality across different regions. This variability can lead to difficulties in standardizing construction practices, which is essential for scaling up affordable housing projects.

Another challenge is the acceptance and perception of these materials. In many cultures, traditional building materials such as concrete and steel are associated with durability and modernity, while sustainable alternatives may be viewed as inferior or outdated. Overcoming these perceptions requires education and awareness campaigns, as well as successful demonstration projects that showcase the viability and benefits of sustainable materials.

Additionally, there are regulatory hurdles related to the use of unconventional building materials. Building codes and standards in many countries are designed with conventional materials in mind and may not accommodate or recognize the performance characteristics of sustainable alternatives. This can lead to delays in approvals, additional costs for testing and certification, and, in some cases, outright bans on the use of these materials.

Technical expertise is also a barrier, as building with sustainable materials often requires specialized knowledge and skills that are not widely available in the construction industry. This can limit the scalability of such practices, particularly in regions where the construction workforce is not trained in these methods.

Finally, the market dynamics of construction materials can pose challenges. The cost of sustainable materials can fluctuate based on supply chain factors, demand, and market maturity. In some cases, sustainable materials may initially be more expensive than conventional options, particularly if they are not produced at scale. This can deter developers from adopting these materials, particularly in the context of affordable housing where cost is a critical concern.

1.6.3 Challenges Hindering the Adoption of Renewable Energy Technologies

The integration of renewable energy technologies into housing developments offers clear benefits in terms of affordability and sustainability, but it also presents significant challenges. One of the primary challenges is the high upfront costs associated with renewable energy systems, such as solar panels, wind turbines, and geothermal systems. These technologies require significant initial investment, which can be a barrier for affordable housing projects that operate under tight budget constraints.

In addition to cost, there are technical and infrastructural challenges related to the installation and maintenance of renewable energy systems. For example, solar panels require adequate sunlight exposure and may not be viable in regions with limited sunshine or where buildings are closely spaced. Wind turbines and geothermal systems similarly depend on specific environmental conditions that may not be present in all locations. Furthermore, maintaining these systems requires specialized knowledge and skills, which may not be readily available in all regions, particularly in developing countries.

Regulatory and policy support is also crucial for the successful adoption of renewable energy in affordable housing. In many regions, the lack of supportive policies, such as subsidies, tax incentives, or feed-in tariffs, makes it difficult for developers to justify the additional costs of renewable energy systems. Furthermore, existing regulations may not accommodate the integration of these systems into housing developments, leading to delays and increased costs.

Grid integration poses another significant challenge. In many areas, the existing electrical grid is not equipped to handle the variable output from renewable energy sources, such as solar or wind power. This can lead to inefficiencies and increased costs, particularly if expensive grid upgrades are required to accommodate the new energy systems.

Lastly, there is the challenge of consumer acceptance and behaviour. For renewable energy systems to be effective, residents must be willing and able to adjust their energy consumption behaviours, such as shifting usage to times when renewable energy is most available. This can be difficult to achieve without comprehensive education and awareness programmes, which require additional resources and time.

1.6.4 Challenges Hindering to Adoption of Offsite Construction

Offsite construction methods, such as modular building and prefabrication, offer significant potential for improving housing affordability by reducing costs and speeding up construction timelines. However, the adoption of these methods is not without challenges. One of the primary challenges is the initial capital investment required to set up offsite manufacturing facilities. These facilities require advanced machinery, skilled labour, and significant infrastructure, all of which demand substantial upfront investment. For many developers, particularly those focused on affordable housing, these costs can be prohibitive.

Another challenge is the logistical complexity of transporting prefabricated components from the factory to the construction site. This process requires careful planning and coordination, as well as access to reliable transportation infrastructure. In regions with poor road networks or other logistical challenges, the cost and complexity of transporting these components can outweigh the benefits of offsite construction.

Regulatory barriers also play a significant role. Building codes and standards are often designed with traditional construction methods in mind and may not readily accommodate offsite construction techniques. This can lead to lengthy approval processes, additional compliance costs, and, in some cases, restrictions on the use of prefabricated components.

Furthermore, there is the challenge of market acceptance. Offsite construction methods are still relatively new in many parts of the world, and there can be resistance from consumers, builders, and even regulators. Concerns about the quality, durability, and aesthetic appeal of prefabricated homes can hinder the adoption of these methods, despite their potential cost savings and efficiency benefits.

Finally, supply chain integration is critical for the success of offsite construction. The production of prefabricated components relies on a well-coordinated supply chain that can deliver materials and components on time and to the required specifications. Disruptions in the supply chain, whether due to material shortages, transportation delays, or other factors, can lead to costly delays and compromise the quality of the final product.

Implementing strategies for providing affordable housing is fraught with challenges across multiple dimensions. Whether through the adoption of zero-energy and carbon buildings, the use of sustainable construction materials, the integration of renewable energy technologies, or the implementation of offsite construction methods, each approach presents its own set of hurdles. Addressing these challenges requires coordinated efforts across industry, government, and communities, along with innovation, investment, and the development of supportive regulatory frameworks. Only through overcoming these challenges can the global housing sector deliver on the promise of affordable, sustainable housing for all.

Digital construction technologies offer one of the most promising solutions to overcoming the myriad challenges associated with providing affordable housing. This will be examined in the ensuing section.

1.7 Adoption of Digital Construction Technologies for Housing Affordability

To effectively address the persistent challenges in delivering affordable housing, the adoption of digital construction technologies stands out as one of the most transformative solutions available today. These technologies offer unprecedented opportunities to enhance technical capacity, streamline processes, and promote effective knowledge sharing across the construction industry. Innovations such as BIM, augmented reality (AR), and blockchain for supply chain transparency have the potential to revolutionize the sector by improving precision, efficiency, and collaboration. BIM, for instance, enables detailed 3D modelling and comprehensive management of information throughout a project's lifecycle, optimizing resource use and significantly reducing waste (Eze et al., 2024). Similarly, AR can enhance on-site execution and training, providing real-time, immersive visualizations that improve accuracy and reduce costly errors. Blockchain technology, on the other hand, offers a robust platform for improving transparency and trust within construction supply chains, ensuring that materials and processes are tracked with precision, thus minimizing delays and cost overruns (Brandín and Abrishami, 2024)

The global construction industry is on the brink of a digital revolution that could drastically mitigate the technical and informational barriers hindering housing productivity and affordability. By integrating digital construction technologies, stakeholders can create a more collaborative, transparent, and efficient approach to housing development. This shift not only requires the adoption of these technologies but also necessitates the establishment of supportive policy and regulatory frameworks that foster innovation and ensure the equitable distribution of benefits across all sectors. Enhancing data collection and analysis capabilities is particularly crucial – not only for understanding housing needs but also for monitoring the impact of digital technologies on improving housing outcomes. By addressing the inefficiencies that have long plagued the construction industry, digital tools can reduce costs, improve quality, and accelerate project timelines, making housing more accessible and affordable on a global scale.

Digital construction technologies are also instrumental in ensuring that other strategies for affordable housing, such as zero-energy buildings and the use of sustainable materials, are implemented efficiently. BIM, for example, serves as a comprehensive platform for acquiring and managing construction data, enabling informed decision-making and overcoming the data scarcity challenges identified by UN-Habitat (2010) as major obstacles to housing provision. Despite being more widely adopted in developed countries, BIM is gradually gaining traction in developing regions, where its benefits are increasingly recognized (Abanda et al., 2014). BIM facilitates active stakeholder involvement, project visualization, and various simulations, which are crucial for integrating sustainable practices into affordable housing projects. However, there remains a gap in BIM components for innovative local materials, limiting the ability to conduct sustainability and lifecycle cost assessments within this digital environment. Developing these components could deliver meaningful, accurate data to support informed decision-making, further advancing housing provision.

The importance of digital construction technologies is underscored by industry reports highlighting the construction sector's lagging productivity growth. For instance, the McKinsey Global Institute reported that productivity growth in construction has stagnated at just 1% annually over the past two decades, largely due to poor data management practices (Barbosa et al., 2017). Improved data management, enabled by digital technologies like BIM, could potentially boost the sector's value by as much as $1.6 trillion, indicating the substantial economic benefits of enhancing information flow and project management (Barbosa et al., 2017). Additionally, a recent publication revealed that 52% (US$280 billion) of rework caused by poor project data and communication globally (Autodesk, 2018) a figure that translates into significant financial losses given the scale of global construction output.

The adoption of digital construction technologies also has a direct impact on reducing project errors and omissions. Studies have shown that BIM can reduce these errors by up to 30%, leading to significant savings in project costs and timelines (Azhar, 2011). This reduction not only decreases direct construction costs but also contributes to lowering the overall market price of new housing developments. Given that rework – often due to poor information management – can account for a significant portion of total project costs, improving data management to minimize rework could substantially reduce the cost burden passed on to homebuyers.

In conclusion, the adoption of digital construction technologies is a critical cross-cutting strategy that enhances the efficiency and effectiveness of housing development, contributing to greater affordability. By leveraging these technologies, the construction industry can overcome longstanding barriers, reduce costs, and improve the quality of housing globally, ensuring that more people have access to sustainable, affordable homes. This book ultimately aims to establish a framework that supports the global adoption and implementation of digital construction technologies, paving the way for a more efficient and equitable housing sector.

1.8 Conclusion

In conclusion, this chapter has laid the foundation for understanding the profound impact that digital construction technologies can have on the global construction industry, particularly in addressing the critical challenges of housing productivity and affordability. By exploring various strategies such as zero-energy buildings, the use of sustainable construction materials, renewable energy adoption, and offsite construction methods, the chapter highlights the multifaceted approaches needed to tackle the housing crisis. However, these strategies are not without their challenges, which range from high initial costs and regulatory barriers to technical expertise and market acceptance issues. Overcoming these hurdles requires a coordinated effort across industries, governments, and communities, as well as the adoption of innovative technologies that can streamline processes and reduce costs.

Digital construction technologies, such as BIM, AR, and blockchain, offer a transformative solution to these challenges by enhancing efficiency, collaboration, and transparency in construction projects. These tools not only improve project outcomes by reducing errors and waste but also play a critical role in making housing more affordable and sustainable. The potential of these technologies to revolutionize the construction industry is immense, and their successful implementation will be key to addressing the global demand for affordable housing. As the book progresses, it will establish a comprehensive framework to support the adoption and integration of these technologies, ensuring that the benefits of digital innovation are realized across the housing sector worldwide.

Chapter 2

Digital Construction Technologies

2.1 Introduction

To provide a comprehensive understanding of the concepts discussed in this book, this chapter will explore several key terms frequently used in the field of construction informatics. The chapter is structured into four main sections to facilitate a clear and organized presentation of the material.

Section 2.2 delves into the definitions of essential terms and concepts that are foundational to the study and application of digital construction technology. This section aims to clarify the terminology that will be used throughout the book, ensuring that readers have a solid grasp of the language and ideas central to this field.

In Section 2.3, we will examine various digital construction technologies in detail. This section provides an in-depth look at the tools, systems, and innovations that are driving advancements in construction practices, highlighting their applications and implications within the industry.

The chapter concludes with Section 2.4, which offers a summary that encapsulates the key points discussed. This final section serves to reinforce the information presented and to provide a cohesive wrap-up of the chapter's content.

2.2 Definitions of Terms and Related Concepts

Digitization versus Digitalization: In the literature, the terms *digitization* and *digitalization* are often used interchangeably, although they are not synonymous. **Digitization** refers to the conversion of information from analogue to digital (binary) form or the process of transforming existing data and processes into a digital format (Gartner, 2022; Ross, 2017). It is also known as digital enablement (Gartner, 2022). For example, converting music from CDs and vinyl records to MP3 files or transforming paper construction documents into digital files stored on a computer (Raaj, 2021).

On the other hand, **digitalization** involves leveraging digital technologies to fundamentally change business operations, enabling them to become digital ventures (Gartner, 2022; Ross, 2017). In other words, digitalization utilizes digital tools to capture and analyse data, enhance decision-making, and create new business models. The broader impact of digitalization across an organization is known as **digital transformation**, which is more of an ongoing process than a fixed outcome.

Digital Transformation: The goal of adopting digital construction technologies is to transform the construction industry or built environment from its underperformance into a more

DOI: 10.1201/9781003476740-2

efficient and productive sector. Thus, although not only common in construction, the term digital transformation is now widely being used in the sector. Although there has been no consensus about the definition of digital transformation, a closer examination of some of the definitions reveals some similarities.

According to Schmarzo (2017), digital transformation is the 'application of digital capabilities to processes, products, and assets to improve efficiency, enhance customer value, manage risk, and uncover new monetization opportunities.' Salesforce (2020) and Guenzi and Habel (2020) refer to digital transformation as 'the process of using digital technologies to create new- or modify existing-business processes, culture, and customer experiences to meet changing business and market requirements.' The Enterprisers Project (undated) defines digital transformation 'as the integration of digital technology into all areas of a business resulting in fundamental changes to how businesses operate and how they deliver value to customers.' Some of the values or benefits of digital transformation include efficient planning and project management (Barnes, 2020), cost savings, productivity and operation efficiencies, improved infrastructure quality, and better environmental performance (Danniotti et al., 2020).

Industry 4.0: Though the term has been popular among researchers and industrialists, there is yet a consensus on its exact definition. The term Industry 4.0 also called 4th Industrial Revolution has been widely used in the scientific literature with uncertainty as to when it was first used. According to Starr et al. (2021), it was first used in 2011. While there is lack of certainty about the first use of Industry 4.0, it was popularized by Klaus Martin Schwab, a German engineer and economist and also founder of the World Economic Forum (Schwab, 2015). Like the lack of consensus about the definition of Industry 4.0, and when the term was first used, the list of technologies associated with it has varied over time with different authors. This is further compounded by the exponential growth of new technologies being invented every year.

According to Rojko (2017), additive manufacturing, rapid manufacturing, adaptive manufacturing, direct digital thread, information systems, physical robots, and software controllers are Industry 4.0. Boston Consulting Group (2020) refers to nine technologies including big data and analytics, autonomous robots, simulation, systems integration, Internet of things, cybersecurity, cloud computing, additive manufacturing, and augmented reality (AR). The Welding Institute list eight technologies that include the Internet of Things (IoT), the industrial Internet of Things (IIoT), cyber-physical systems (CPS), smart manufacture, smart factories, cloud computing, cognitive computing, and artificial intelligence (AI) (TWI, 2022).

Smart Cities 4.0: First of all, the smart city is defined as a municipality that uses information and communication technologies (ICT) to increase operational efficiency, share information with the public, and improve both the quality of government services and citizen welfare (Shea and Burns, 2020). Therefore, Smart City 4.0 is defined as an emerging city model that leverages the collective power of engaged and digitally connected citizens to enhance the quality of urban life and ensure sustainable urban development. Technically, Smart City 4.0 is a sub-concept of Smart City. Cities 4.0 is used in the literature to mean Smart Cities 4.0 (Wray, 2021)

Construction 4.0: Presently, there is no global consensus as to what really is Construction 4.0. Some view it as a simple instantiation or subclass of Industry 4.0 whereby ubiquitous technologies are used for real-time decision-making in the construction sector (Craveiro et al., 2019; Forgues, 2019; Schober, 2016). Others are more generic in their views and define it as the digitization of the industry and the automation of its processes (Carey, 2021). A more comprehensive definition encompassing most definitions is that of the European Industry Construction Federation (FIEC). FIEC defines Construction 4.0 as 'the counterpart of Industry 4.0 in the

Architecture, Engineering and Construction (AEC) industry and it refers to the digitalization of the construction industry' (FIEC, 2022).

Logistics 4.0 or Smart Logistics: Logistics 4.0 involves system-based planning and control of the physical movement of goods and its associated information flow from the source to the final destination (Bag et al., 2020). Logistics 4.0 uses front-end and base technologies (Frank et al., 2019) and provides logistics solutions to customers by meeting their demands without an increase in costs. Logistics 4.0 mainly uses technologies like IoT, CPS, big data analytics, and cloud computing (Winkelhaus and Grosse, 2020).

Supply Chain 4.0: According to Alicke et al. (2016), Supply Chain 4.0 is

> the application of the Internet of Things, the use of advanced robotics, and the application of advanced analytics of big data in supply chain management: place sensors in everything, create networks everywhere, automate anything, and analyze everything to significantly improve performance and customer satisfaction.

Procurement 4.0: Procurement 4.0 is the integration of Industry 4.0-related technologies into the procurement process, making tasks automated, information gathering, and communication more effective, while establishing interconnected networks.

Teleoperation: This is also known as remote operation or telerobotics. The technology enables the control and manipulation of devices, machines, or robots from a distance. It allows an operator, located at a remote location, to interact with and control a robotic system or other machinery without direct physical contact. The primary goal of teleoperation is to extend human capabilities and overcome limitations posed by geographical distances, hazardous environments, or other challenging conditions. Teleoperation has been widely adopted as a promising way to enhance worker safety in extreme and hazardous construction workplaces (Lee et al., 2022).

Construction Automation: There is still a lack of consensus on a clear definition of construction automation. Scholars have proposed various definitions with regard to 'construction automation,' 'construction robotics' or 'construction robot,' 'construction automation and robotics (CAR).' Skibniewski (1992) defined construction automation as the engineering of any construction process using teleoperated, numerically controlled, semi-autonomous, or autonomous equipment, while construction robotics is defined as advanced construction equipment capable of being teleoperated, acquisition, and analysis of sensory data and the execution of numerically controlled or autonomous tasks. Mahbub (2008) defined CAR as the use of self-control mechanical and electronic machinery with intelligent control mechanisms to conduct construction tasks and operations automatically. The used terms and definitions overlap in use, covering systems from mechanical machinery manipulated by the human, to semi-automated or automated devices with remote control, to autonomous robots with more sensors and automated control features. For the purpose of this study, construction automation and robotics is considered in a boarder sense as including a wide spectrum of machinery and applications for automating construction-related processes in the whole lifecycle of the project. The aims of construction automation are many, including but not limited to, reducing jobsite injuries, decreasing activity completion times, and assisting with quality control and quality assurance (McKinsey & Company, 2019). Some systems may be fielded as a direct response to increasing skilled labour shortages in some countries (Vernon and Hughes, 2017). Opponents claim that increased automation may lead to less construction jobs and that software leaves heavy equipment vulnerable to hackers (Carter, 2019). Bock and Linner (2015) categorized construction automation into five main groups, namely, robot-oriented design, robotic industrialization, construction robots, site automation, and ambient robotics.

Technology: Technology can be the knowledge of techniques, processes, and the like, or it can be embedded in machines to allow for operation without detailed knowledge of their workings. Systems (e.g., machines) applying technology by taking an input, changing it according to the system's use, and then producing an outcome are referred to as technology systems or technological systems. The International Institute of Applied System Analysis splits technology into three main categories.

- Hardware which refers to Manufactured objects (artifacts). For example, a physical personal computer on a desk.
- Software which refers to knowledge required to design, manufacture, and use technology hardware. In order words, software is a collection of instructions and data that tell the computer how to work. For example, Revit is a software installed on a computer that tells it how to represent a wall when instructed by a human expert.
- Orgware which refers to institutional settings and rules for the generation of technological knowledge and for the use of technologies.

ConTech: ConTech, short for Construction Technology, refers to the application of various technological solutions and innovations within the construction industry to improve productivity, efficiency, safety, and overall project management. It encompasses a wide range of digital tools and technologies that are reshaping how construction projects are planned, executed, and monitored. ConTech aims to address the challenges faced by the construction industry, such as labour shortages, project delays, safety concerns, and the need for increased productivity. It refers to technologies that can be used to we plan, design, and build structures, as well as the manufacture and installation of their components (DesignBuildings Wiki, 2021).

PropTech or Prop-Tech: PropTech, short for Property Technology, refers to a category of technology that streamlines and optimizes various aspects of the real estate industry. It encompasses a wide range of residential and commercial real estate software and platforms that aim to improve systemic inefficiencies, digitize transactions, and enhance decision-making processes throughout the real estate lifecycle. PropTech connects people, technology, and information to simplify how real estate is bought, sold, managed, operated, researched, rented, marketed, and invested in.

It is the application of information technology and platform economics to real estate markets (Shaw, 2019).

RealTech is commonly used in the United States to cover real estate technology. It refers to various technologies used in the real estate sector to create environments through strategic procedures and activities. RealTech companies focus on modernizing traditional real estate systems and processes using cutting-edge software, hardware, and core technologies to eliminate inefficiencies and outdated practices.

CREtech: This stands for **c**ommercial **r**eal **e**state **tech**nology or corporate real estate technology.

It refers to the application of various technological innovations and solutions within the commercial real estate industry. It encompasses a wide range of digital tools and technologies designed to streamline and optimize various aspects of commercial real estate operations, transactions, and management. CREtech aims to enhance efficiency, data analysis, decision-making, and tenant experiences in the commercial real estate sector. It refers to digital tools a company uses to efficiently run its facilities and the spaces within those buildings. The term is often used synonymously with PropTech though PropTech refers to both commercial and residential technology.

Real EsTech: This refers to technologies used in the real estate industry that affect the built environment through either strategic innovation or product enhancement. This contrasts with PropTech which is more aligned to specific real estate purposes and transactional practices. Other similar terms have been used including Building 4.0 (The Construction Index, 2022) and Civils 4.0 (Pitcher, 2018).

2.3 Types of Digital Construction Technologies

The use of digital technologies is the bedrock underpinning the conceptualization of digital transformation (Olanipekun and Sutrisna, 2021). The last decade has experienced a surge in the use of digital technologies in the construction industry. Nowadays, digital technology is much talked about in the industry. In spite of the technology being widely spoken about there is no clear and agreed definition of what is 'digital technology' really is.

This lack of consensus is inherent in the words that make up 'digital construction.' The word 'digital' is quite confusing and could mean different things to different people. The word 'construction' is misleading too. Is it about the phase of actual on-site realization of a project or a whole lifecycle process from start-to-end?

However, Mills (2016a) makes an effort to define digital construction as the use and application of digital tools to improve the process of delivering and operating the built environment. As argued by Mills (2016b), digital construction might be simple tools that make communication easier and reduce the need for travel (messaging tools like Slack and WhatsApp or video conferencing systems like Skype and Google Hangouts). It could be the improvement or automation of a manufacturing process, advancements in plant or materials, cloud-based computing and filing systems, or software applications for use in both delivery and operation/facilities management. Or it could be considerably more advanced than that; things at the cutting-edge like unmanned aerial vehicles (UAVs, sometimes referred to as 'drones') for site scanning or inspections, 3D and 4D printing, robotics, or AI. According to UK Connect (2021), the top 10 types of digital construction technologies impacting the industry: Big Data, AI and machine learning (ML), the IoT, Robotics & Drones, 5D and Wi-Fi6, building information modelling (BIM), virtual reality (VR)/AR, 3D Printing, Mobile and Cloud Technologies, and Blockchain.

In the literature, there are so many digital construction technologies, which at times can be overwhelming to researchers and interested users. The various technologies have been classified into four categories which include digital data, automation system, digital access, and connectivity (Dallasega et al., 2018; Heusler and Kadija, 2018; Olanipekun and Sutrisna, 2021). Digital technologies generate data when used; for example, wearable sensors and smart meters are used as a collection point of *digital data* in construction (Craveiroa et al., 2019). The *automation* systems use digital technologies to create self-organizing systems such as robots for lifting objects on sites (Berlak et al., 2020) and blockchain for executable payment to contractors (Li et al., 2019). *Digital access* refers mobile access to internet networks to execute solutions in real time such as data analytics and processing to make real-time decisions or make future predictions (Berger, 2016; Maskuriy et al., 2019). *Connectivity or network* encompasses the linking and synchronizing separate activities such as 3D model development and energy-use simulation in the BIM platform (Keskin et al., 2020) or linking the physical-to-digital-to-physical in construction using sensors, cloud computing, IoT, AR, and VR (Craveiroa et al., 2019).

2.3.1 *Robots*

A robot is defined as, 'a machine capable of carrying out a complex series of actions automatically, especially one programmable by a computer'; the robots can be both fully and semi-automated. According to Essers and Vaneker (2013), the use of industrial robots is important in performing repetitive tasks while using these machines will enable tasks to be performed cost effectively and accurately over longer periods of time. This is due to the reduced wastage through less human error and the ability for machines to work over an extended time without breaks (unlike a standard 8-hour day with breaks for a human).

Within the construction industry, there is a wide range of robotic use with a varied allocated purpose. Construction robots can be split into two sub-categories, manufacturing robots and on-site labour-based robots. Manufacturing robots, often in the shape of mechanical arms or production lines, and on-site robots used for construction activities. Manufacturing robots are commonly used on production lines for material production, for example, the QT10-15 Block Making Machine that requires an input of pure concrete and is able to fully cast blocks in an automated way. These types of robots have benefits, as described by Bikas et al. (2016), including high accuracy, speed, strength, reliability, and repeat capability. This enables the production line to be more efficient in terms of cost and time. However, Kousi et al. (2019) challenge the benefits and suggest these robots are still not fully developed with issues in their flexibility and issues with reconfiguration. This is due to three reasons, 'the rigidity of the used stationary robots, the use of fixed and product model dedicated equipment, and the use of fixed robot control logic.'

The on-site-based robots vary widely in use; they include those that are vehicular based to assist workers and those that assist directly with the construction activities, like bricklaying and rebar tying. There are different types of vehicular robots with smaller, autonomous 'rovers' that are able to follow workers and carry their tools or materials and larger vehicles like the Volvo HX models that are autonomous dumper trucks. The activity-based construction robots are used to assist or complete tasks, often laborious tasks or those with repetitive movements, for example, the Semi-Automated Mason (SAM 100), a bricklaying robot which only requires an assistant mason to apply the mortar finish while the TyBot replaces labourers in the process of tying rebar. This is specifically associated with bridge construction, which is considered high risk by the National Institute for Occupational Health and Safety declaring that 'the frequency and duration of the hand and wrist motions are associated with increased risk of a hand-wrist WMSDs' (work-related musculoskeletal disorders).

Another robot being implemented in the industry in the SAM 100 is a bricklaying robot, designed and developed by Construction Robotics, with an automated arm being the key feature. The robot works collaboratively with a mason who loads the robot with premixed mortar and bricks while the mason also removes excess mortar and ensures the required finish is applied. The robot itself has two conveyor belt brick feeds, two propane tanks which assist the arm movements, a brick-measuring tool so the robot knows the size of brick it is laying and a safety wheel which, as long as there is always a supervising mason, would allow the robot to work around the clock. Detailed architectural drawings are uploaded to the robots' computer system which allows the robot to know how many bricks to lay and in what positions, allowing for openings.

In the construction industry, robots can be classified into the following groups: robot dog (Kim et al., 2022), articulated, cartesian, collaborative, and UAH robots (Fu, 2025):

Robot Dog: Robotic pets are artificially intelligent machines that are made to resemble actual pets. Examples of common robotic pets are Joy for All (by Hasbro) Companion Pets, Zoomer

Interactive Kittens and Puppies, PARO Robot Seals by Intelligent Systems Co., and Robot Dog.

Articulated Robots: Articulated robots are a type of robot that feature rotary joints, allowing them to have multiple degrees of freedom and a wide range of motion. These robots typically consist of a series of interconnected segments or links, resembling a human arm, with each joint providing rotational movement. The number of joints or axes can vary, ranging from two-jointed structures to complex systems with 10 or more interacting joints.

Articulated robots are widely used in industrial settings and are considered the most common type of industrial robot today. They offer great versatility and precision, closely mimicking the movement of a human arm. These robots can perform various tasks and applications, including material handling, welding, assembly, and machine tending.

Cartesian Robots: Cartesian robots, also known as linear robots or gantry robots, are a type of industrial robot that operates based on Cartesian coordinate geometry. These robots utilize linear axes for movement, typically along the X, Y, and Z axes, which are perpendicular to each other. The primary characteristic of Cartesian robots is their ability to move in a straight line rather than rotate. They consist of an overhead structure that controls horizontal motion and a vertical robotic arm or actuator that enables vertical movement. Cartesian robots can be designed to operate in two-dimensional (x–y axes) or three-dimensional (x–y–z axes) space, depending on the specific configuration and application.

Collaborative Robots (Cobots): Supporting workers but still requiring human supervision or operation. As a quality of interaction, fluency in joint action refers to a highly coordinated cooperation between the human and the robot who form a well-synchronized synergy-driven team, adapt to each other, and perform together to complete the task (Paliga and Pollak, 2021).

Unmanned Aerial Vehicles (UAVs) – drones, becoming more and more popular in construction industry. An UAV, commonly known as a drone, is an aircraft that operates without any human pilot, crew, or passengers on board. These vehicles were initially developed for military purposes to perform missions that were considered too dull, dirty, or dangerous for humans. However, their use has expanded to numerous non-military applications as control technologies have improved and costs have decreased.

UAVs or drones have found applications in various fields such as aerial photography, precision agriculture, environmental monitoring, forest fire monitoring, river monitoring, infrastructure inspections, policing and surveillance, product deliveries, entertainment, and drone racing. They offer advantages in terms of manoeuvrability, range, and endurance compared to manned aircraft, as they are not burdened by crew and life-support systems.

2.3.2 D-Laser Scanners

Some important concepts will have to be defined before defining 3D laser scanners. One such concept is Lidar (also known as LIDAR, LiDAR, and LADAR). LIDAR is a method for determining ranges (variable distance) by targeting an object or a surface with a laser and measuring the time for the reflected light to return to the receiver. Its full meaning is 'light detection and ranging' (National Ocean Science, 2021) or 'laser imaging, detection, and ranging' (Taylor, 2019).

As Wang et al. (2018) state, LiDAR is capable of providing highly accurate, high-resolution data in a three-dimensional way with the use of multiple beams. LiDAR is a special type of sensor that captures the surroundings and presents them in a range of dots that mark how far the light beam has ranged through timing how long it takes to return to the sensor. The dots can then be exported and presented in a 3D mesh which displays the surrounding area (LiDAR, 2019). Wang et al. (2018) explain how these data can be used to create 'occupancy grid maps' which lay out the surroundings into regular grids with the associated data stored here.

It can also be used to make digital 3D representations of areas on the Earth's surface and ocean bottom of the intertidal and near coastal zone by varying the wavelength of light. It has terrestrial, airborne, and mobile applications. It is sometimes called 3D laser scanning, a special combination of 3D scanning and laser scanning (Shan and Toth, 2018). 3D Laser scanning is a modern technique used to accurately represent physical objects in the virtual world. It is a non-contact, non-destructive technology that digitally captures the shape of physical objects using a line of laser light. 3D laser scanners create 'point clouds' of data from the surface of an object (LaserDesign, 2022). In other words, 3D laser scanning is a way to capture a physical object's exact size and shape into the computer world as a digital 3-dimensional representation.

***Terrestrial 3D Laser Scanners*:** A terrestrial 3D laser scanner is a ground-based version of the airborne LIDAR frequently used for terrain and landscape mapping. It has a stand or tripod on which the scanner is set.

Handheld 3D Laser Scanners: A handheld 3D laser scanner is a scanner for capturing large objects, such as aeroplanes, wind turbines, rooms, or buildings. As the name implies, it is held in the hand during the scanning process.

Wearable 3D Laser Scanners: Wearable 3D laser scanners are laser scanning technologies that can be worn on the human body. Some wearable scanners can be mounted on the shoulders called shoulder-mounted 3D laser scanners and others on the wrist called wearable wristband 3D laser scanners.

Mobile 3D Laser Scanners: Mobile 3D laser scanners are laser scanners that are mounted on mobile objects during operations. Some examples of mobile objects could be robots and drones.

Mobile Apps 3D Laser Scanners: The mobile technology industry is expanding everyday with new products. There are now many applications (Apps) in smart phones', iPads or other smart devices' Apps stores. Some of the Apps can be used for 3D laser scanning. They are called Mobile Apps 3D laser scanners.

A summary of the different type of 3D laser scanners and their vendors/manufacturers is presented in Table 2.1.

2.3.3 3D Printers

The concept of 3D printing will be explained using additive building manufacturing (ABM). The term 'additive manufacturing' (AM) refers to the set of additive manufacturing production processes starting from digital models, as opposed to traditional subtractive techniques (machining by chip removal, cutting, and drilling) (Conner et al., 2014). ABM involved the use of large robots on a building site to 3D Print (extrude) building materials to construct buildings (AABM, 2016). Aerial ABM refers to aerial robotic construction systems that enable aerial robots to 3D print building structures autonomously (AABM, 2016).

In the literature, there are about 10 different types of 3D printing technologies depending on the printing techniques. These are fused deposition modelling (FDM), stereolithography, digital light processing (DLP), selective laser sintering (SLS), material jetting (MJ), drop on demand

Table 2.1 Types 3D Laser scanners

Main category	*Types*	*Applications*	*References*
Terrestrial	Trimble X8	The Trimble X12 quickly captures features for land title surveys, buildings, roads, intersections, site improvements, encroaching elements, and complex structures	https://geospatial.trimble.com/products-and-solutions/trimble-x12
	RIEGL VZ-6000	Its high accuracy, long-range capabilities, and comprehensive data acquisition features make it a versatile tool for various industries requiring precise 3D measurements and modelling.	http://www.riegl.com/
	Stonex X300 New	This technology is designed to provide accurate data every time and for any project.	https://geo-matching.com/products/x300-new
	Smart Max Geosystems VS1000	It can be used in engineering surveying, urban building surveying, topographic mapping, mining, deformation monitoring, factory/large structures/pipeline design, aircraft and ships manufacturing, road and rail construction, tunnelling, bridge reconstruction, water conservancy, airport and port engineering, and other fields	https://www.manualslib.com/products/Smart-Max-Geosystems-Vs1000-5142813.html
Wearable	NavVis VLX	NavVis VLX is an end-to-end reality capture solution for laser scanning, surveyors, and engineers.	https://www.navvis.com/
	Leica Pegasus	This can be used in digitization of infrastructures, planning, and resources of cities.	https://leica-geosystems.com/en-gb/products/mobile-mapping-systems/capture-platforms/leica-pegasus-backpack
Handheld	GeoSLAM Floorplans	It delivers accurate, fast, and repeatable floorplans with an easy-to-understand workflow, helping businesses to unlock hidden value in their commercial real estate assets.	https://geoslam.com/solutions/floorplans/
	GeoSLAM Construction Progress	Automatically generate clear progress reports with visualizations that show you the build progress made on site. It's used by construction companies to track and record progress against a 3D model or a previous scan to improve efficiency, save money, create unbiased progress reports, and provide valuable data to help resolve any disputes that may arise.	https://geoslam.com/solutions/geoslam-construction-progress/
	GeoSLAM Hub & Draw	It transforms 3D point cloud data into actionable information and valuable deliverables.	https://geoslam.com/solutions/scan-to-bim/
	FARO Freestyle 2	Fast, photorealistic 3D reality capture for scanning construction projects	https://www.faro.com/en/Products/Hardware/Freestyle-2-Handheld-Scanner

	Leica-BLK2GO	This can be used to capture an entire building, structure, or other large environment in minutes.	https://www.sccssurvey.co.uk/blk2go.html
Mobile 3D laser scanner	Spot	Robot-mounted for on-site surveying	https://www.bostondynamics.com/products/spot
	Faro Trek	Robot-mounted for on-site surveying	https://www.faro.com/en/Products/Hardware/Trek-3D-Laser-Scanning-Integration
	Trimble's Turnkey Robotic Scanning Solution	Robot-mounted for on-site surveying	https://fieldtech.trimble.com/en/product/spot
	Leica RTC360 3D laser scanner integration with Spot robot	Robot-mounted for on-site surveying	https://leica-geosystems.com/en-gb/about-us/news-room/news-overview/2021/02/2021_02_22_boston_dynamics
	Leica BLK2FLY	Drone-mounted site and development surveying	https://leica-geosystems.com/en-gb/products/laser-scanners/autonomous-reality-capture/blk2fly
	YellowScan	Drone-mounted site and development surveying	https://www.yellowscan-lidar.com/
Mobile Apps 3D Laser scanner	Scandy Pro	Scandy Pro allows users to turn their smartphones or tablets equipped with depth-sensing cameras into portable 3D scanners. This capability enables individuals in various industries, such as architecture, engineering, product design, and art, to create accurate 3D models of objects and spaces quickly and easily.	https://www.scandy.co/#Intro
	AR Plan 3D	AR Plan 3D is a room scanner for scanning spaces.	https://apps.apple.com/us/app/ar-plan-3d-camera-to-plan/id1459846158
	3D Live Scanner	This can be used in providing high-quality building services models.	https://play.google.com/store/apps/details?id=com.lvonasek.arcore3dscanner&hl=en_GB&gl=US

Table 2.2 Types of 3D Printers

Types	*Description*	*References*
MakerBot	It prints carbon fibre-reinforced nylon and other engineering-grade composite parts with three-dimensional strength and high accuracy	https://www.tritech3d.co.uk/manufacturer/makerbot-3d-printers/
Ultimaker	The Ultimaker S5 Pro Bundle boosts productivity, flexibility, and confidence to unlock demanding applications with an extended range of materials.	https://shop3d.ca/products/ultimaker-s5-pro-bundle?srsltid=AfmBOorNDGDK6YJKXanIK9bHNIaUOJ2_2pQI8WJi13U8Skx1SGiWDtRW
FlashForge	Most types (e.g., Adventurer 4) are equipped with flexible build plates to achieve strong adhesion and fast heating, which helps ensure a high success rate.	https://www.flashforgeshop.com/
CreatBot	It provides solutions for just about every application from concept to prototype and high-performance material direct production.	https://www.creatbot.com/
XYZPrinting	This printer can produce high-quality 3D colour print	https://pro.xyzprinting.com/en-GB/home
3D Systems	These are industrial 3D printers that can design, test and produce metal parts through 3D printing, not possible with standard manufacturing.	https://www.3dsystems.com/
Beeverycreative	It beautifully designed, elegant and portable 3D printer for personal uses.	https://all3dp.com/topic/beeverycreative/

(DOD), sand binder jetting, metal binder jetting, direct metal laser sintering (DMLS), selective laser melting (SLM), and electron beam melting (EBM) (MB, 2022) (Table 2.2).

Also, there exists powerful software application specifically designed for additive manufacturing and 3D printing workflows. The software provides tools for design, optimization, simulation, and preparation of 3D models, especially those intended for printing with industrial-grade 3D printers. Some examples included in Table 2.3.

The 3D printers have also got mobile app equivalents that can be downloaded from the Mobile App Store or Google Store. Some examples are Creality Cloud, BuildBee Easy 3D, GCodePrintr, EasyPrint 3D, and AstroPrint.

2.3.4 *Blockchains*

A blockchain is a distributed database or ledger that is shared among multiple nodes in a computer network. It is best known for its crucial role in cryptocurrency systems, where it maintains a secure and decentralized record of transactions. However, the applications of blockchain extend beyond cryptocurrencies and can be used to make data in any industry immutable, meaning it cannot be altered.

The structure and access of data in a blockchain differ from traditional databases. A blockchain consists of blocks linked together through cryptographic hashes. Each block contains a hash of the previous block, a timestamp, and transaction data. This chain of blocks forms a secure and tamper-resistant record of transactions.

Table 2.3 3D Printing software

Fusion 360	Fusion 360 to create and then edit your objects for 3D printing	https://all3dp.com/2/fusion-360-3d-printing-tutorial-for-beginners/
Netfabb	Netfabb has been specifically developed for additive manufacturing, design, and simulation to streamline workflows and reduce build errors before fabrication	https://www.3dnatives.com/en/netfabb-software-241120205/#!
SolidWorks	It supports 3D printing of models that contain surface bodies and graphics bodies only if they form a watertight manifold volume.	https://help.solidworks.com/2021/English/SolidWorks/sldworks/c_Print3D.htm
Onshape	It is a CAD software that is specifically designed for 3D printing workflows	https://www.onshape.com/en/resource-center/ebooks/is-your-cad-software-the-best-choice-for-3d-printing
Tinkercad	It allows users to create models that are compatible with 3D printing	https://www.3dnatives.com/en/tinkercad-all-you-need-to-know-120320204/
Solid Edge	It uses synchronous technology, a method of 3D modelling which gives a user the freedom to change between direct and parametric modelling.	https://www.sculpteo.com/en/3d-learning-hub/3d-printing-software/solidworks-vs-solidedge/
Blender	It supports 3D printing modelling and file formats since 2002.	https://www.blender.org/
Ultimaker Cura	It is an open-source slicing application for 3D printers	
Siemens NX	It provides a single software environment that can design, optimize and prepare builds for 3D printing technology.	https://europac3d.com/reverse-engineering-software/siemens-nx/

2.3.5 Drones

Drones can be used in the delivery of on-site projects in many ways. Firstly, they can provide easy access to high-rise or hard to reach parts of infrastructure, such as underneath bridges, remote railways, and hydropower dams. Drones can also be used to monitor the progress of work, deliver materials on-site, and land surveying, and they have been found to be a viable solution to the safety issues that exist when working at height.

Using drones during infrastructure construction is a less time-consuming, more cost-effective, and safer method of construction as there is little – to no contact between individuals, meaning the use of drones will lessen the spread of the virus (Fan and Saadeghvaziri, 2019) (Table 2.4).

2.3.6 The Automatic Identification and Data Capture (AIDC) ('aka Automatic Identification,' 'Auto-ID,' and 'Automatic Data Capture')

AIDC refers to a sequence of data capture prior to identification (Furness, 2000). The aspect of identification may refer to the identification of animate or inanimate objects, data, locations,

Table 2.4 Types of drones

Type	*Description*	*Reference*
Multi-rotor	A multi-rotor drone is a type of UAV or drone that uses multiple rotors to achieve lift and propulsion.	https://www.auav.com.au/articles/drone-types/#2
Fixed-wing	A fixed-wing drone has one rigid wing that is designed to look and work like an aeroplane, providing the lift rather than vertical lift rotors	https://www.auav.com.au/articles/drone-types/#2
Single-rotor	A single-rotor drone, also known as a single-rotor helicopter or single-rotor UAV, is a type of unmanned aircraft that relies on a single large rotor for lift and propulsion.	https://www.auav.com.au/articles/drone-types/#2
Fixed-wing hybrid VTOL	A fixed-wing hybrid VTOL (vertical take-off and landing) drone is a type of unmanned aerial vehicle (UAV) that combines the characteristics of both fixed-wing aircraft and multi-rotor drones. This hybrid design allows the drone to take off and land vertically like a multi-rotor drone while also enabling it to transition into forward flight like a conventional fixed-wing aircraft.	https://www.auav.com.au/articles/drone-types/#2

and events. The data aspect can be a simple data string that allows reference or linking to data or information stored elsewhere (so-called licence plate concept), or a portable data file, each with attributes that extend beyond the process of identification.

2.3.7 *Configurator*

A configuration system can be defined as an IT system capable of 'combining well-defined building blocks governed by rules and constraints into a product' (Andreasen, 1987). The implementation of a configuration system thus implies a well-defined modular structure of the product. The basis for a product configuration system is a product model containing 'all relevant information concerning a given product during the lifecycle' (Krause et al., 1993). Product configuration as described by Hvam et al. (2008) is an effective way of structuring products composed of standard parts, and product configuration is also a method of presenting products to customers. An example of a configurator is Configurator 360 developed by Autodesk.

2.3.8 *Internet of Things*

The IoT refers to the network of physical objects or 'things' embedded with sensors, software, and connectivity, allowing them to collect and exchange data with other devices and systems over the internet. In other words, it is a conceptual framework that leverages on the availability of heterogeneous devices and interconnection solutions, as well as augmented physical objects providing a shared information base on a global scale, to support the design of applications involving at the same virtual level both people and representation of objects (Atzori et al., 2017).

2.3.9 *Virtual Reality (VR)*

VR refers to a computer-generated simulation in which a person can interact within an artificial three-dimensional environment using electronic devices, such as special goggles with a screen or gloves fitted with sensors. This technology creates a simulated environment that can mimic or completely differ from the real world, allowing users to immerse themselves in a digital landscape. VR systems are designed to fully immerse the user in the virtual environment, making them feel as if they are experiencing the simulated reality firsthand. This immersion is achieved through a combination of visual, auditory, and sometimes tactile feedback. VR has applications across various fields, including entertainment (gaming and virtual tours), education and training (simulated training environments for medical, military, and aviation personnel), design (architecture and engineering visualization), and therapy (treatment of phobias and anxiety disorders). The key aspect of VR is its ability to create a sense of presence, where users feel as though they are physically present in the virtual environment.

VR is a technology which allows a user to interact with a computer-simulated environment, whether that environment is a simulation of the real world or an imaginary world (Mandal, 2013).

2.3.10 *Augmented Reality (AR)*

AR is a technology that superimposes digital information – such as images, videos, or sound – onto the real world through the use of devices like smartphones, tablets, or specialized AR glasses. Unlike VR, which creates a completely artificial environment, AR enhances the user's perception of the real world by adding digital elements to it. This blending of digital content with the physical environment allows users to interact with and perceive both the real and virtual components simultaneously.

AR applications are diverse, ranging from entertainment and gaming to practical uses in education, healthcare, retail, and construction. In construction, for example, AR can overlay a building's design plans onto the physical jobsite, allowing for real-time comparison and adjustments. In education, it can bring textbooks to life, providing interactive 3D models that students can explore. The immersive nature of AR offers a unique, enhanced experience that can improve learning, facilitate complex processes, and enhance understanding and engagement in various fields.

AR is a technology that supplements the real world with virtual (computer-generated) objects/elements and enables users to interact with the virtual elements (Bekele and Champion, 2019).

2.3.11 *Mixed Reality (MR)*

Mixed reality (MR) blends elements from both the real and virtual environments to create a real–virtual environment that enhances our perception of both environments (Bekele and Champion, 2019). This environment enables interaction between users, virtual elements, and the real world, leading to a user–reality–virtuality interaction and relationship space. Unlike AR, where the real environment is dominant, MR does not allow one environment to dominate the other, instead both environments benefit from each other's elements.

AR overlays digital content onto the real world, enhancing one's current perception of reality. VR creates a fully immersive digital environment, isolating the user from the physical world. MR blends real and virtual worlds, allowing interaction with and manipulation of physical and digital objects together. Each offers unique experiences and applications, from enhancing real-world views with digital overlays (AR), to immersing users in entirely digital environments (VR), to blending both for interactive experiences (MR).

2.3.12 *Metaverse*

The Metaverse refers to a virtual, shared, and interconnected universe that encompasses all virtual worlds, AR spaces, and digital environments. It is a collective and immersive virtual space where users can interact, socialize, create, and engage with each other and digital content. The concept of the Metaverse is heavily influenced by science fiction and has gained significant traction with advancements in technology, VR, AR, and the internet. Some selected AR, VR, and MR platforms have been presented in Table 2.5, while their associated headsets have been presented in Table 2.6.

The list below is just a small selection of drones that are suitable for construction-related tasks; however, you should definitely search around as there are always new models being released and one of them may be more suitable for your needs.

2.3.13 *Digital Versus Physical Twins: What Is the Difference?*

In the literature, different views as to what constitutes a digital twin have been examined. Despite the term becoming more widely used in the construction industry, there is no agreed definition of what it is (Jowett, 2019).

According to Digital Built Britain, a digital twin is 'a realistic digital representation of assets, processes or systems in the built or natural environment' (CDBB, 2018). Similarly, in Designing Building Wiki, 'digital twin' refers to a virtual model or replica of assets, processes, systems, and other entities (BIM Wiki, 2023).

A range of definitions of digital twin has been examined in Onesait Cloud Platform (2018). These are discussed in the ensuing paragraphs.

Similar to the definition on the Design Building Wiki website, a digital twin is a digital replica of a physical asset, process, system across its lifecycle. Digital twin which is a virtual representation of the physical world is a very similar to simulation that uses real-time data to promote understanding, identify problems, develop new opportunities, and plan for the future.

> A digital twin is an exact digital replica of a product, process or service. This living model creates a thread between the physical and digital world. IoT-connected objects are replicated digitally, enabling simulations, testing, modelling and monitoring based on the data collected by IoT sensors. Like everything in the realm of IoT, data is the primary driver, and most invaluable output, of digital twins. The sharing and analysis of digital twin data empowers companies to make decisions, which directly impact their key performance indicators.

Summarily, a **digital twin is a digital representation of a real-world asset or system that does not act as a substitute for the physical object of the system it represents, but as a replica of it, allowing the communication (testing, monitoring, control) of this physical device without having to be stuck to it. As noted in Centre for Digital Built Britain, 'what distinguishes the "digital twin" from any other digital models is its connection to the "physical twin"'** (CDBB, 2018).

Table 2.5 AR, VR, and MR platforms

Main category	*Type*	*Application*	*Reference*
AR	HoloSite	An integrated cloud-based solution plugging into our AR construction headset – The Atom.	https://www.xyzreality.com/product
	VIMaec	VIMaec is an AR platform that converts BIM models into immersive AR experiences. It allows stakeholders to experience BIM designs in real-world environments, supporting design communication and decision-making.	https://www.vimaec.com/
	Trimble Connect	It is a cloud-based collaboration platform for construction and design teams. It includes AR functionality, enabling users to view and interact with BIM models overlaid in the real environment using AR devices.	https://connect.trimble.com/
	HoloBuilder	It is an AR construction documentation platform that supports BIM model integration. It allows users to overlay BIM models onto physical spaces, providing a comprehensive view of design and construction progress	https://www.holobuilder.com/
VR	VREX	It enables virtual collaboration in real-time inside the model for construction projects.	https://www.vrex.no/
	WVS	It is a cloud platform for interactive creators to develop and ship virtual worlds and applications on real-time game engines.	https://wevr.com/
	IrisVR Prospect	IrisVR Prospect is a VR collaboration platform designed specifically for architecture, engineering, and construction (AEC) professionals. It converts 3D models from popular design software (e.g., Revit, Rhino, SketchUp) into VR experiences, allowing stakeholders to immerse themselves in the design and identify potential issues.	https://irisvr.com/
	Unity Reflect	Experience immersive and collaborative real-time 3D environments, including in AR and VR, for better decision-making across your entire project lifecycle.	https://unity.com/products/unity-reflect
	The Wild	The Wild is a virtual collaboration platform that enables real-time VR meetings and design reviews. It integrates with BIM and 3D modelling software, allowing teams to explore and edit design ideas together in a shared virtual space.	https://thewild.com/
	Prospect by SketchUp (or SketchUp Viewer for VR)	Prospect by SketchUp was a VR extension for SketchUp that allowed users to view their SketchUp models in virtual reality environments. With Prospect, users could experience their 3D designs in a more immersive way, providing a better understanding of spatial relationships and design elements.	https://irisvr.com/integrations/sketchup
MR	Theorem-XR	Visualization, design review, and factory layout (defining, reviewing, editing, and validating shop floor layouts.)	https://www.theorem.com/
	Upskill – Skylight	Upskill Skylight is an enterprise-focused MR platform that delivers AR experiences to industrial workers. It provides hands-free access to contextual information, instructions, and remote assistance for field technicians.	https://ar4industry.be/wiki/upskill-skylight/

Table 2.6 Types of headsets

Types	*Market product*	*References*
AR	Magic Leap 2	https://www.magicleap.com/magic-leap-2
VR	Samsung Gear	https://www.samsung.com/uk/wearables/gear-vr-r325/SM-R325NZVABTU/
	WorldViz	https://www.worldviz.com/
	Meta Quest 2	https://www.worldviz.com/post/how-to-connect-your-oculus-quest-2-to-vizard-and-vizible-using-oculus-link
MR	HoloLens	https://www.microsoft.com/en-us/hololens
	Varjo XR-3:	https://varjo.com/
	Microsoft HoloLens 2 Smart Glasses (NJX-00002)	https://www.bechtle.com/gb/shop/microsoft-hololens-2-smart-glasses–4342628-01–p

The connection between Digital Twins and BIM lies in their complementary roles in the construction and operation of buildings and infrastructure:

Data Integration: BIM models can serve as the foundation for creating Digital Twins. By integrating BIM data with live sensor data and real-time information, Digital Twins can offer insights into the performance, maintenance, and operational aspects of a building or infrastructure, enhancing decision-making throughout the asset's lifecycle.

Lifecycle Management: Digital Twins extend the usefulness of BIM beyond construction by providing ongoing monitoring, analysis, and optimization capabilities. They allow stakeholders to visualize, monitor, and optimize operational assets, processes, and resources using live data, which is crucial for efficient management and oversight throughout the lifecycle of a construction project or facility.

In summary, Digital Twins and BIM are closely connected technologies, with BIM serving as a foundation for creating Digital Twins and Digital Twins extending the capabilities of BIM into the operational phase, enabling better decision-making, efficiency, and collaboration throughout the asset's lifecycle.

2.4 Conclusions

As we conclude our exploration of digital construction technologies in Chapter 2, we reflect on the transformative potential these innovations hold for the construction industry. The shift from traditional methods to digitalized processes, through the adoption of technologies such as BIM, 3D printing, drones, and smart sensors, promises a future where construction projects are more efficient, sustainable, and adaptable to the changing needs of society.

The discussions highlighted the importance of understanding the nuances between digitization and digitalization, and how the construction sector is moving beyond mere digitization towards a more integrated and holistic digital transformation. This journey involves not just the adoption of new technologies but also a cultural shift within organizations, emphasizing the need for upskilling, collaboration, and change management.

Moreover, the chapter illustrated the role of Construction 4.0 as a subset of the broader Industry 4.0 movement, showcasing how interconnected, intelligent systems can revolutionize project management, operations, and execution. With examples ranging from the use of drones for site surveying to the application of AI for predictive maintenance, it's clear that the digital age offers unprecedented opportunities for innovation in construction.

However, challenges such as data privacy, cybersecurity, and the digital divide among firms pose significant hurdles to the widespread adoption of these technologies. To navigate these challenges, the construction industry must foster an environment of continuous learning, regulatory compliance, and ethical technology use.

As we look to the future, it's evident that the journey of digital transformation in construction is ongoing. The next chapter will delve deeper into BIM, a cornerstone of digital construction technologies. We will explore BIM's principles, applications, and impact on project lifecycle management, highlighting how it represents a paradigm shift in how buildings are designed, constructed, and maintained. Through detailed examination, readers will gain insights into leveraging BIM for enhanced collaboration, efficiency, and innovation in their projects.

In the upcoming chapter, we will explore the intricate details of BIM, a technology that has become synonymous with the digital transformation of the construction industry. This chapter will dissect the components, processes, and benefits of BIM, offering readers a comprehensive understanding of how this technology can be applied to achieve greater efficiencies, reduced costs, and improved outcomes in construction projects. From conceptualization to operation, we will navigate through the BIM lifecycle, highlighting case studies and best practices that underscore its significance in the modern construction landscape.

Chapter 3

Building Information Modelling

3.1 Introduction

In the dynamic and ever-evolving world of modern construction, where efficiency, collaboration, and precision are of paramount importance, building information modelling (BIM) has established itself as a transformative force reshaping the industry. This chapter embarks on a thorough exploration of BIM, delving deep into its multifaceted nature. We will cover everything from foundational definitions to its vital role as a process, as well as the groundbreaking technologies that drive its widespread adoption and success.

To provide a structured and comprehensive understanding of BIM, the remainder of the chapter is divided into three key sections. Section 3.2 presents a critical appraisal of the definition of BIM, examining how it is understood and applied across different contexts within the construction industry. This section aims to clarify the concept by dissecting various perspectives and interpretations, helping readers appreciate the depth and breadth of BIM as more than just a tool, but as an integral process within construction.

In Section 3.3, the focus shifts to the classification and analysis of emerging BIM technologies. This section explores the latest innovations that are enhancing the capabilities of BIM, enabling more efficient and precise construction processes. We will discuss the tools, platforms, and methodologies that are pushing the boundaries of what BIM can achieve, offering insights into how these technologies are shaping the future of construction.

The chapter concludes in Section 3.4, where we summarize the key points discussed, tying together the concepts and technologies introduced earlier.

3.2 What Is BIM?

While the phonetics of BIM may great, its acronym, although also easy to pronounce, has been and is very misleading. Each letter of the acronym will be examined to reveal how it has been rightly or wrongly misinterpreted.

The '**B**' in **B**IM is often taken to mean 'BIM is for buildings' only. This is not true as BIM can be implemented in managing any construction or built environment information. BIM can be used for infrastructure development, bridges, oil and gas, water, power, and roads and transport management. The B which stands for 'Building' should be used as a verb. That is to say the act of 'building' or constructing a construction object. In other words, 'B' for building refers to the realization of any construction artefact.

The '**I**' of B**I**M represents information or data that is to be embedded into the building model. In fact, this is where BIM significantly deviates from conventional computer-aided design

DOI: 10.1201/9781003476740-3

systems. The core of BIM is about managing information or data (the 'I' of BIM) that has been embedded in the building model in a way that is useful for the project. This is best explained by drawing an analogy with Jess Greenwood who said 'Data is the new oil. It's only useful when it is refined!' This is the non-geometric data of a building model.

The '**M**' of BI**M** represents the geometric model of the construction project or object. While this in itself is not wrong, it however gives the impression BIM is all about geometric three-dimensional (3D) models. Generating 3D models for construction projects is not new at all. For generations, CAD systems have been used in generating 3D models. Thus, the real intended meanings of the letters of the BIM acronym can be summarized as follows:

B
Building – So long as it is a verb
I
Information – It is all about the information/data
M
Modelling – Maybe… too much focus on graphics

So, in a nutshell, BIM is about managing information, and the information consists of the geometric as well as non-geometric data. Managing, exchanging, and sharing construction information is the core principle of BIM. Hence, in the recent literature, an additional M is now appearing in the acronym or M simply being replaced by other terms other than 'modelling.' Synonyms like Building Information Management, BIM & Management (BIM&M), and Better Information Mobility (BIM) are now common in the literature.

What really is BIM? Three formal definitions will be examined in the ensuing paragraphs.

The UK Construction Industry Council (CIC) defines BIM as …

> an innovative and collaborative way of working that is underpinned by digital technologies which support more efficient methods of designing, creating and maintaining the built environment.

Mott MacDonald defines BIM as…

> a coordinated set of processes, supported by technology, that add value by creating, managing and sharing the properties of an asset throughout its lifecycle. Models incorporate graphic, physical, commercial, environmental and operational data.

Lee et al. (2006) define BIM as: the 'process' of generating and managing building information in an interoperable and reusable way; and a BIM system is a system or a set of systems that 'enables' users to integrate and reuse building information and domain knowledge through the lifecycle of a building.

What emerges from the aforementioned definitions is the fact that information management can be supported through the use of digital technologies in a BIM process. The role of digital technologies is quite important as it allows professionals to virtually explore building or construction project models before realization on site. As a result, synonyms such as 'virtual models,' 'build before you build,' 'build it twice' (once virtually and once in reality), and 'digital twin' are now being used as equivalents to BIM.

The collaborative working principle of BIM can be summarized in equation 3.1.

$$\begin{aligned}\textbf{BIM} = \textbf{Process} + \textbf{Technology} + \textbf{People} + \textbf{Policy} + \textbf{Standards}\\ + \textbf{Legal Implications} + \textbf{Security}\end{aligned} \tag{3.1}$$

Process: BIM as a process is better encapsulated by Lee et al. (2006), who define BIM as '… the "process" of generating and managing building information in an interoperable and reusable way….' Three keywords underpin Lee et al. (2006) definition: generate, manage, and interoperable. Generating information is a process. Generation process could be designing and producing schedules and quantities using a BIM authoring tool such as Revit. If the quantities generated is to be used for a given application, then it has to be moved through a process often called 'export' to a specified file format. Once the export activity is completed, the BIM application software now reads or 'import' the file for its use. The 'export' and 'import' actions are processes. The concept of collaboration is key the UK Construction Industry Council (CIC) definition of BIM. Collaboration is a working practice whereby individuals work together for a common purpose to achieve business benefits. Collaboration enables individuals to work together to achieve a defined and common business purpose. Collaboration can take two forms: Firstly, synchronous, where everyone interacts in real time, as in online meetings, through instant messaging, or via Skype. Secondly, asynchronous, where the interaction can be time-shifted, as when uploading documents or annotations to shared workspaces, or making contributions to a wiki. In the act of collaboration, data/information is generated and shared among project team members for use in different applications depending on their tasks in the BIM project. To ensure a structure approach of collaboration, and efficient data/information management, a collaborative environment is needed. For example, using any common data environment can be effective in bringing teams together, facilitating the generation, management, and sharing of construction information.

Technology: The process of sharing construction information can be supported with emerging technologies. Here, technology is considered as an enabler, it is not BIM, and BIM is not technology. In the literature and in practice, so many technologies exist that can support the BIM workflow. The different technologies can be categories into software and hardware.

Software: The different software can be classified into design, application, and management software.

The design software often called BIM authoring tools or software are those used for designing 3D models. Two most common examples are Revit and ArchiCAD. The application software consists of those used for designing common construction applications. For example, iTwoCostX is used for developing cost plans; Navisworks and Synchro are used for producing project programmes or 4D BIM. The third category consists of software used in managing construction information. This category is often called common data environments. Common examples of common data environments are BIM360, 4Project, and BIMXtra. For an extensive review of these software, readers are advised to Abanda et al. (2015) and Abanda et al. (2018).

Hardware: The different hardware can be classified into two groups. The first relates to computers on which the different software can be installed. In order to implement BIM, it is imperative to match the requirements of the different software to be installed and the specification of the computer. Most vendors or manufacturers usually include the specification of the computer on their websites. Without this, a chosen computer may lack the memory to accommodate a given BIM software. In some cases, when the memory of the computer is low, the software may be too slow and can easily crash when loaded with heavy BIM models. The second relates to smart devices used for sharing construction information. These include iPad, iPhone, and all smart phones that can be used in sharing construction information, especially with site managers.

***Related Technologies*:** Currently, the industry is witnessing a surge in emerging technologies that together with BIM can improve the process of construction information capture, management and capturing. Some examples are robots, drones, 3D laser scanners, and 3D-printers. A major consideration for integration of data from these technologies with different BIM software is interoperability. Can the data generated by these technologies be read or exported into BIM software?

***People*:** Process and technology may be fully developed with a great potential in sharing information, but if people do not change attitudes towards embracing BIM, it will still not go anywhere.

New jobs or roles have emerged as a result of BIM implementation in practice. The roles have been attributed to various names. The most common BIM-related roles in job adverts are BIM coordinator, Document controller, Built asset security manager, CAD manager, BIM manager, BIM modeller, Interface manager, Information Manager, GSL champion, Architectural BIM lead, GSL lead, Revit technician, Task information manager, Project delivery manager, BIM technician, BIM engineer, Model originator, Task team manager, and Model authors.

***Policies*:** Due to the nature of man or people, policies may be needed to drive the uptake of BIM. The policies can be mandatory or voluntary. Different countries have rolled out different policies at different periods. A summary of some of them has been examined in United BIM (2022) presented in the ensuing paragraphs.

The United Kingdom: In April 2016, the government mandated BIM Level 2 on all the public projects. UK's construction industry is one of the most technologically advanced and digitized industries in the world.

The United States: BIM is not mandated across all the states yet, but it is expected to grow significantly. Wisconsin state implemented and made it mandatory to implement BIM for public projects if equal or above the **total budget of $5 million.**

France: France is on the roadmap to digitalize its construction sector, and it has also developed BIM standards for infrastructure project works. In 2014, they initiated by developing 5,00,000 houses using BIM. They have also allocated funds of around Euros 20 million to enhance the construction industry digitally fully. In the future**, implementation of BIM will be mandatory in public projects**.

Scandinavian Region: Norway, Denmark, Finland, and Sweden were the first ones to get their hands-on BIM adoption. **Mandates for public projects are in place in these countries**. They are one of the global leaders in BIM adoption and implementation.

Singapore: The construction authority of Singapore has implemented BIM on all the public projects since 2015. The building and construction authority, CORNET, had implemented BIM e-submission in 2015. They have made it mandatory to adopt BIM, for all the projects greater than 5,000 sq. mts. Singapore's construction industry has shown productivity because of the implementation of BIM.

United Arab Emirates: In UAE, many of the famous buildings were delivered using BIM and this practice has been on the rise before the mandate was passed by the government, which was in the year 2014. It is mandatory for all the projects that are 40 stories and higher, or 30,000 sq. ft. and larger.

China: The adoption of BIM in China has a significant increase since 2016. Chinese AEC professionals and several organizations have integrated a high level of BIM **adoption**

policies for growth and digitization. BIM has become a crucial element, and it is now being used for most of their projects. Even the government of China is taking a keen interest in BIM adoption policies. They have not yet mandated BIM, but its use is positively encouraged.

Australia: The adoption of BIM remains fragmented in Australia as there is no consistent approach to it. Few private players are implementing BIM. However, it lacks consistency. In 2016, the government urged to create a smart infrastructure task force to successfully implement BIM on all public projects above $50 million

Standards: With regard to conventional working, a standard establishes the methodology for managing the production, distribution, and quality of construction information, including that generated by CAD systems, using a disciplined process for collaboration and a specified naming policy. In a like manner, standards are required in a BIM process. There are five main standards regulating BIM practice. These are ISO 19650-1, ISO 19650-2, ISO 19650-3, ISO 19650-4, and ISO 19650-5.

Legal Implications of BIM: Lastly, man is the hallmark of capitalism, trying to gain the maximum benefit and pushing the risk away to others down the supply chain. Also, working in a collaborative environment means traceability of actions can be very challenging. Should an incident occur as a result of the contributions from various stakeholders, who will take the responsibly or be liable? Issues related to ownership, are considered under the legal aspects (legal implications of BIM).

Security: According to ISO 19650-5: 2020, security is defined as the state of relative freedom from threat or harm caused by deliberate, unwanted, hostile, or malicious acts. Working on BIM-compliant projects, information is exchanged between stakeholders through the use of machines or common data environment platforms such as Autodesk Construction Cloud. Therefore, the security of information about assets is crucial particularly for critical projects such as that of government buildings, prisons, and power plants. Inappropriate distribution and loss of sensitive information may potentially lead to physical security threats, financial loss, and loss of trust and reputation (Das et al., 2021). According to ISO 19650-5:2020, sensitive information is defined as:

'Information, the loss, misuse or modification of which or unauthorized access to, can:

- adversely affect the privacy, security or safety of an individual or individuals
- compromise intellectual property or trade secrets of an organization
- cause commercial or economic harm to an organization or country and/or
- jeopardize the security, internal and foreign affairs of a nation.'

Consequently, stakeholders working on BIM projects should be security-minded, i.e., understanding and routinely applying appropriate and proportionate security measures in any business situation so as to deter and/or disrupt hostile, malicious, fraudulent, and criminal behaviours or activities.

BIM: The UK Construction Industry Council defines BIM as … 'an innovative and collaborative way of working that is underpinned by digital technologies which support more efficient methods of designing, creating and maintaining the built environment.'

City Information Modelling (CIM): CIM generally involves building a 3D city model that connects with BIM and other contextual data source or analysis tool of various city components, including buildings, roads and public spaces (open data), streetlights (sensors/IoT), and even people on the street (social media) (Hisham, 2018).

Geographical Information Systems (GIS): GIS can be defined as a computerized system for capture, storage, retrieval, analysis, and display of spatial data describing the land attributes and environmental features for a given geographic region, by using modem information technology (Thurgood and Bethel, 2003).

3.3 BIM Software Systems

BIM software systems are tools that facilitate the creation, management, and collaboration of building information models in the AEC industry. They provide functionalities for designing, coordinating, visualizing, and simulating building projects, ultimately improving project outcomes and efficiency. Professor Charles Eastman trained as an architect and worked in Computer Science at Carnegie Melon University developed the early BIM software systems including:

- Building Description System (BDS)
  - BDS was a database that contains graphical data, sortable data that allow the user to retrieve information categorically by attributes including material type and supplier
- Building Design Advisor, in 1993 developed in Lawrence Berkeley National Laboratory

While developments were going on in the USA, two programming geniuses emerged in the Soviet Block: Leonid Raiz and Gábor Bojár (Physicist)

The technological timeline by LetsBuild (2023):

1957 – Pronto: It is considered the first commercial CAM software, marking the beginning of computer-aided manufacturing processes.

1963 – Sketchpad: Sketchpad introduced the concept of a graphical user interface (GUI) for CAD, allowing users to interact with digital objects in a more intuitive way.

1975 – BDS: BDS was an early attempt to digitize building information, laying groundwork for later developments in BIM.

1977 – Graphical Language for Interactive Design (GLIDE): GLIDE represented an advancement in interactive CAD systems, focusing on graphical input and output.

1982 – 2D CAD: The adoption of 2D CAD systems became widespread, significantly improving drafting efficiency and accuracy.

1984 – Radar CH: Radar CH contributed to the development of architectural design software, though specific details about its impact are less commonly noted.

1985 – Vectorworks: Vectorworks began as a general CAD programme but has evolved to include specific features for architecture, landscape, and entertainment design industries.

1986 – Really Universal Computer-Aided Production System (RUCAPS): RUCAPS is one of the early systems that used BIM concepts, allowing for the integration of information in building models.

1987 – ArchiCAD: ArchiCAD was a pioneer in BIM, offering architects the ability to work on building models in three dimensions with data attached.

1988 – Pro/ENGINEER: Pro/ENGINEER (now PTC Creo) introduced the concept of feature-based, parametric solid modelling in mechanical CAD.

1992 – BIM: The term BIM was officially coined, defining the process of generating and managing building data during its life cycle.

1993 – Building Design Advisor: Building Design Advisor offered tools for integrating and visualizing building design decisions.

1994 – miniCAD: miniCAD (later known as Vectorworks) continued to evolve as a versatile CAD platform for various design disciplines.

1995 – International Foundation Class (IFC): The IFC file format was introduced to facilitate interoperability in the building industry, allowing different software applications to exchange data.

1997 – ArchiCAD's Teamwork: ArchiCAD's Teamwork feature revolutionized collaborative work in BIM, enabling multiple team members to work on the same project simultaneously.

1999 – Onuma: Onuma is a BIM and cloud-based platform, emphasizing planning and design in an online environment.

2000 – Revit: Revit was launched as a revolutionary BIM-centric architectural design and documentation software.

2001 – NavisWorks: NavisWorks (later acquired by Autodesk) became known for its project review software, enabling 3D model visualization and collaboration.

2002 – Autodesk buys Revit: Autodesk's acquisition of Revit marked a significant investment in BIM by one of the industry's leading software companies.

2003 – Generative Components: Generative Components introduced computational design into the mainstream, allowing for complex shape generation based on parametric rules.

2004 – Revit 6 update: The Revit 6 update continued to advance the capabilities of BIM software, enhancing usability and performance.

2006 – Digital Project: Digital Project, a suite of architecture and engineering tools, offered advanced BIM capabilities, particularly for complex geometries.

2007 – Autodesk buys NavisWorks: The acquisition of NavisWorks by Autodesk further consolidated their portfolio of construction and design collaboration tools.

2008 – Parametricist Manifesto: The Parametricist Manifesto advocated for the use of parametric design and computational methods in architecture, influencing the direction of software development.

2012 – Formit: Formit introduced a more intuitive, touch-based approach to preliminary 3D design and modelling, integrating with Revit and other BIM tools.

3.3.1 BIM Authoring Tools

BIM authoring tools are software applications used in the field of BIM to create, manage, and manipulate digital models of buildings and their components. These tools enable architects, designers, engineers, contractors, and other stakeholders to collaborate and work together throughout the entire lifecycle of a building project.

BIM authoring tools organize the model by objects, attributes, and relationships, allowing users to create a virtual 3D geometric model that represents the physical aspects of the building accurately. These tools are primarily used in architectural, structural, plant, and infrastructural design disciplines. They facilitate the creation of informed 3D models and support the coordination of various contributions related to the project, such as diagrams, tables, and other graphics

- Revit (Autodesk)
- ARCHICAD (Graphisoft)
- Allplan (Nemetschek)
- ArCADia BIM
- AECOsim Building Designer (Bentley Systems)

3.3.2 *Object Libraries*

BIM object libraries are collections of digital objects that can be used in BIM projects. These libraries contain preconfigured 3D models and data of various building components and products, such as furniture, fixtures, equipment, materials, and systems. BIM object libraries provide designers, architects, engineers, and other professionals with a convenient way to access and incorporate specific products into their BIM models, saving time and ensuring accuracy in the design process.

There are several platforms and websites that offer BIM object libraries. These libraries are often categorized by product type, manufacturer, or industry-specific requirements. Users can browse through the libraries, search for specific products, and download the BIM objects in formats compatible with their BIM software.

Some well-known BIM object library platforms include BIMobject, NBS National BIM Library, and curated libraries offered by BIMobject itself. These platforms provide a wide range of BIM objects from various manufacturers, both generic and manufacturer-specific, which can be downloaded for free.

- BIMObject
- NBS
- BIMStore
- ARCAT
- Revit City

3.3.3 *Visual Programming Language (VPL)*

VPL is a programming language that allows users to create programmes by manipulating programme elements graphically instead of using traditional textual coding. It utilizes visual expressions, graphical symbols, and spatial arrangements to represent programme logic and flow. Instead of writing lines of code, users work with graphical elements, such as blocks, icons, diagrams, and flowcharts, to construct the programme.

The main goal of VPLs is to make programming more accessible and intuitive, especially for novices or users without a programming background. VPLs often provide a visual interface where users can drag and drop elements, connect them together, and configure their properties to create a programme. This visual approach aims to reduce syntactic errors, improve understanding of programming primitives, and support the exploration of programme behaviour in different scenarios.

- Dynamo
- Grasshopper
- Karamba
- Blueprint
- Marionette for Vectorworks
- Houdini
- Bentley Systems GenerativeComponents
- Sverchok for Blender
- Karamba3D
- BIMscript
- NodeBox

3.3.4 4D BIM Tools

4D BIM refers to the incorporation of scheduling or time-related information into a BIM model. It is a process that enables the visualization and simulation of the construction sequence and timeline of a project. It involves linking the 3D model components or assemblies with the project delivery timeline, allowing for resource scheduling, quantity tracking, and project phasing. 4D simulations provide valuable communication and visualization tools, helping identify and resolve potential bottlenecks during construction.

- Think Project
- Navisworks
- Synchro
- Asta Power Project
- Bentley Navigator
- Fuzor

3.3.5 5D BIM Tools

5D BIM tools refer to software applications or platforms that incorporate the dimensions and cost in addition to the three spatial dimensions (width, length, and depth) provided by 3D BIM models. The integration of cost dimensions enhances the capabilities of BIM by enabling project teams to visualize, simulate, analyse, and manage construction schedules and costs more effectively. 5D BIM involves integrating design data with estimating, scheduling, and costing information. This integration enables the generation of Bills of Quantities, derivation of productivity rates and labour costs, and the ability to link cost estimates to the BIM model. By incorporating cost information, project teams can assess the financial implications of design changes and make informed decisions. Some selected 5D BIM software are:

- iTwoCostX
- Vico
- CubiCost
- Dprofiler
- Assembly Systems
- Kreo

3.3.6 Clash Detection Software

BIM clash detection software is a tool used to identify, inspect, and report clashes or collisions between elements within a structure's 3D model. Clash detection is crucial for coordinating and resolving conflicts that may arise when different disciplines, such as architectural, structural, mechanical, electrical, and plumbing, develop their respective models independently. Some selected clash detection software are:

- Tekla BIMsight
- Navisworks
- Solibri Model Checker

- usBIM.clash
- MagiCAD
- Fuzor
- Trimble Connect

3.3.7 BIM for Structural Analysis and Design

BIM for structural analysis and design software refers to software applications that utilize BIM technology specifically tailored for structural engineering tasks. These software tools assist structural engineers in designing, analysing, and documenting structural systems, as well as integrating design with fabrication processes. Some selected clash detection software is:

- Robot Structure
- STAAD Pro
- CYPECAD
- RAM (Bentley)
- Tekla Structural Designer
- SkyCiv Structural 3D

3.3.8 6D BIM Tools (BIM Energy Efficiency Software)

In the context of BIM, the sixth dimension, also known as 6D BIM, involves analysing the sustainable performance of construction projects. Some selected 6D BIM software is:

- Integrated Environmental Solutions (IES)
- openBuildings Energy Simulator (Bentley)
- Sefaira
- EcoDesigner (Gaphisoft)
- CypeCAD MEP(CYPE)

3.3.9 7D BIM Tools (BIM Software Facility Management)

The seventh dimension, also known as 7D BIM, extends the scope of BIM to include facility management information. It involves the integration of building performance data, maintenance schedules, asset management, and other relevant information to support the efficient operation and maintenance of a building throughout its lifecycle. With 7D BIM tools, stakeholders such as facility managers, building owners, and operators can access comprehensive information about a building's components, systems, and maintenance requirements. These tools enable better decision-making, as they provide insights into the long-term viability, performance, and costs associated with different design options. Some selected 7D BIM software is:

- Ecodomus
- ArchiBus
- ArchiFM (Graphsoft)
- Onuma
- FM: Systems

3.3.10 Common Data Environments (CDE)

A CDE is a digital information platform or integrated IT solution that serves as a centralized repository for collecting, managing, and disseminating project and asset information throughout the lifecycle of a construction project. The CDE is a collaborative environment where all stakeholders involved in the project can work on and share information in a structured and controlled manner. The primary purpose of a CDE is to enable effective information management and ensure that the right people have access to the right information at the right time. It helps streamline workflows, improve collaboration, and reduce errors and inefficiencies that can arise from fragmented data and disjointed processes. By implementing a CDE, construction teams can enhance coordination, maintain data integrity, and improve project outcomes. Some selected CDEs are:

- Autodesk Construction Cloud
- Asite
- planBIM
- Plannerly
- Aconex
- Procore
- BIMXtra
- ProjectWise

3.3.11 Mobile BIM

Mobile BIM, or mobile building information modelling, refers to the use of mobile devices and applications in the field of construction and building information management. BIM itself stands for BIM, which is a holistic process of creating and managing digital representations of physical and functional characteristics of built assets.

Mobile BIM utilizes mobile devices such as smartphones and tablets, along with specialized software applications, to enable real-time communication and access to building information while on-site or in remote locations. It allows construction professionals, including architects, engineers, contractors, and project stakeholders, to interact with and utilize BIM data directly from their mobile devices, enhancing collaboration and decision-making in the construction process.

- planBIM
- Autodesk Construction Cloud
- BIMx
- StreamBIM
- Zutec BIM
- Trimble Connect

3.3.12 Modelling and Rendering Software

Modelling and rendering software are tools used in the field of computer graphics to create and visualize 3D models and scenes. These software applications enable artists, architects, designers, and other professionals to design, simulate, and render realistic and detailed representations of objects, environments, and animations.

Modelling software provides the tools and features necessary to create 3D models from scratch or by manipulating existing geometric shapes. It allows users to define the shape, structure, and appearance of objects, including their dimensions, textures, and materials. Modelling software can range from basic tools that focus on simple shapes and objects to advanced applications capable of creating complex and highly detailed models. Some popular modelling software include:

- Maya
- Rhinoceros
- Maxwell Render (Maxwell)
- ArtLantis (Abvent)
- SketchUp (Timble)
- 3D StudioMax
- Lumion
- Cinema 4D(Maxon)
- Enscape

3.3.13 Microsoft

Microsoft does not have a dedicated BIM software of its own. However, Microsoft has been involved in providing cloud-based services and collaboration tools that can support BIM workflows. Here are some Microsoft tools and services that are commonly used in conjunction with BIM processes:

- MS Excel
- Microsoft Azure
- Power BI
- Microsoft 365
- Tableau

3.3.14 BIM Plugins

BIM plugins, also known as Revit plugins or add-ons, are software extensions that enhance the functionality and capabilities of BIM software, specifically Autodesk Revit. These plugins are designed to automate tasks, provide additional features, and improve productivity for BIM users, including architects, engineers, and designers.

BIM plugins can perform a wide range of functions and address various needs within the BIM workflow. They can simplify complex tasks, improve efficiency, and enhance the overall user experience. Some common functionalities offered by BIM plugins include:

- DiRoots.One
  - SheetLink
  - FamilyReviser
  - OneFilter
  - TableGen
  - SheetGen

    - ReOrdering
    - ParaManager
    - PointKit
- BIM Interoperability tools
- BIMVision
- Proving Ground
- RTVTools
- Family Browser
- RushForth tools
- Ideate Software Bundle

3.4 Conclusions

In the exploration of BIM and its technologies within the construction industry, we have uncovered the transformative power of digitalization in shaping the future of construction projects. BIM, as detailed in this chapter, stands not just as a technological advancement but as a paradigm shift towards a more integrated, collaborative, and efficient approach to construction.

The comprehensive examination of BIM reveals its multifaceted role in enhancing project management, design, construction, and maintenance phases. By embedding detailed information into digital models, BIM facilitates a level of precision and foresight previously unattainable, enabling stakeholders to make informed decisions at every project stage. The exploration of software systems, authoring tools, and emerging technologies further underscores the breadth and depth of BIM's impact, highlighting its capacity to drive improvements in cost, time, quality, and sustainability.

However, the journey towards fully realizing BIM's potential is not without its challenges. Issues such as interoperability, data privacy, and the need for standardized practices emerge as critical hurdles. These challenges underscore the importance of continued innovation, collaboration, and a commitment to standards that promote seamless information exchange and integration across diverse platforms and stakeholders.

As we look ahead, the focus shifts towards common concepts, system integration, and interoperability in the next chapter. This forthcoming discussion is vital, as it addresses the foundational elements that enable the effective use of BIM technologies. Integration and interoperability are crucial for the seamless exchange and use of information, ensuring that the myriad tools and systems employed in BIM workflows can communicate and function together efficiently. By delving into these concepts, we aim to provide a roadmap for overcoming the barriers to effective information exchange, paving the way for a more connected, efficient, and innovative construction industry.

The next chapter delves into the essential aspects of common concepts, system integration, and interoperability in the realm of BIM technologies. This discussion is pivotal, as it addresses the core challenges and solutions associated with the seamless exchange of information across different systems and platforms within the construction industry. We will explore the theoretical underpinnings of system integration and interoperability, practical approaches for achieving seamless data exchange, and the standards and protocols that support these objectives. Through detailed insights and case studies, readers will gain a comprehensive understanding of how to effectively implement and leverage BIM technologies for enhanced collaboration, efficiency, and innovation in their construction projects.

Chapter 4

Common Concepts, System Integration, and Interoperability

4.1 Introduction

In today's fast-paced and interconnected world, technology is at the heart of nearly every aspect of our lives. From everyday mobile applications to sophisticated enterprise systems, the successful operation of these technological advancements relies heavily on their ability to communicate and collaborate effectively. This chapter embarks on an exploration of the fundamental concepts of system integration and interoperability, highlighting their crucial role in creating a cohesive and efficient technological environment.

To ensure a clear and structured understanding, the remainder of the chapter is organized into five key sections. Section 4.2 introduces and discusses common building information modelling (BIM) concepts, providing the foundational knowledge necessary to grasp the broader topics that follow. In Section 4.3, the focus shifts to system integration, examining how different technologies and platforms are brought together to work as a unified system. Section 4.4 delves into system interoperability, exploring how different systems interact and share information seamlessly to enhance efficiency and reduce barriers.

In Section 4.5, the openBIM paradigm is examined in detail, showcasing its significance in promoting openness, collaboration, and transparency in the construction industry. Finally, the chapter concludes in Section 4.6, where the key points are summarized, tying together the various concepts discussed throughout the chapter to reinforce their importance in the modern technological landscape.

4.2 Common BIM Concepts

4.2.1 BIM Maturity Models

Level 0: In its simplest form, Level 0 effectively means no collaboration. 2D CAD drafting only is utilized, mainly for production information. The output and distribution are via paper or electronic prints, or a mixture of both. The majority of the industry is already well ahead of this now.

Level 1: At this level, information is managed using CAD in 2D or 3D. Typically, 2D CAD is used for drafting of statutory approval documentation and production information, while 3D CAD is primarily used for conceptual design and visualization. CAD standards are managed to BS 1192:2007, and electronic sharing of data is carried out from a common data environment (CDE), often managed by the contractor.

DOI: 10.1201/9781003476740-4

At this level, there is:

- very limited or no data attached to the model
- no federation

Level 2: Is the creation of a managed 3D environment with data attached, but created in separate, distinct discipline models. These separate models may originate with the client, architect, structural engineer, building services engineer, contractors, etc. BIM Level 2 or the 'federated model' as it is sometimes referred requires each participant to develop their own models which are then shared with the project model, with appropriate audit trails in approved environments such as CDE.

At this level, there is:

- federation, but each participant works separately and then drops their model in a common data environment
- no real-time and simultaneous collaboration

Level 3: This is a fully integrated and collaborative process enabled by 'web services' and compliant with emerging IFC standards. It involves the fuller integration with the parties to a building project working simultaneously on a centrally hosted model.

At this level, there is:

- real-time and simultaneous collaboration

In the literature, different diagrams have been used to illustrate the concept of BIM maturity. A popular example is presented Figure 4.1.

Level 4: Level 4 BIM introduces the concepts of improved social outcomes and wellbeing. This is understood as 'better perceived social outcomes that can improve the lives and wellbeing of people which can be successfully engineered through effective requirements management and integrated data processing.'

With emerges from the aforementioned maturity models is the gradual capacity requirements to be attained by organizations at the different levels in the practice. The recent BS ISO 19650 does not adopt a triangular maturity model. The ISO 19650 presents a new BIM maturity model (Figure 4.2) that supersedes the most popular and widely adopted version of Figure 4.1

From Figure 4.2, the new model does not use the 'level' terminology, it uses stages, and they run from 1 to 3. Comparatively, BIM Level 0 has been left out from the new maturity model. Also, BIM Level 2 will from henceforth be called 'BIM according to ISO 19650' or in practical terms designing according to 'BIM according to ISO 19650.'

Stage 2 maturity is also identified as 'BIM according to the ISO 19650 series.' This is where a mixture of manual and automated information management processes is used to generate a federated information model. The information model includes all information containers delivered by task teams in relation to an asset or a project. The ISO 19650 series has application mainly at Stage 2 maturity, but also can be partly be applied at Stages 1 and 3.

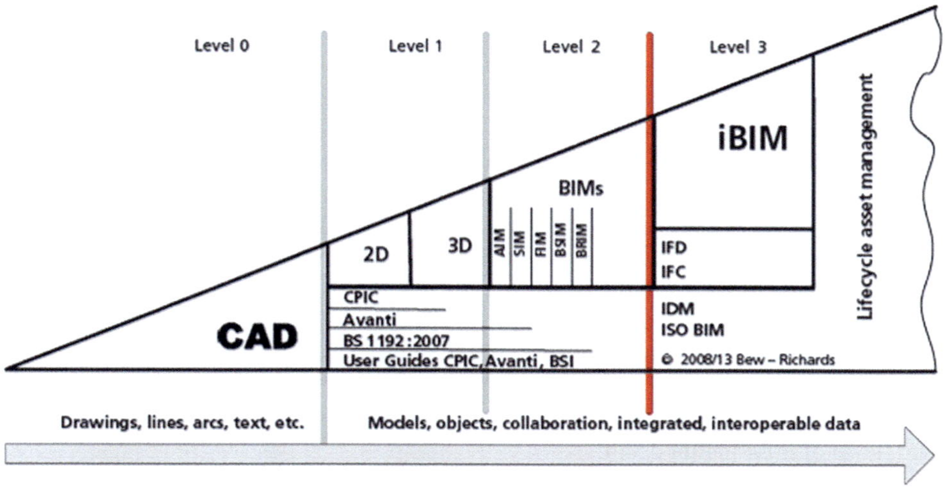

Figure 4.1 UK BIM maturity model in 2D (Source: PD 19650-0:2019 (2019))

[Permission to reproduce extracts from British Standards is granted by BSI. British Standards can be obtained in PDF or hard copy formats from BSI Knowledge: https://knowledge.bsigroup.com *or by contacting BSI Customer Services for hardcopies only: Tel: +44 (0)20 8996 9001, Email:* cservices@bsigroup.com.]

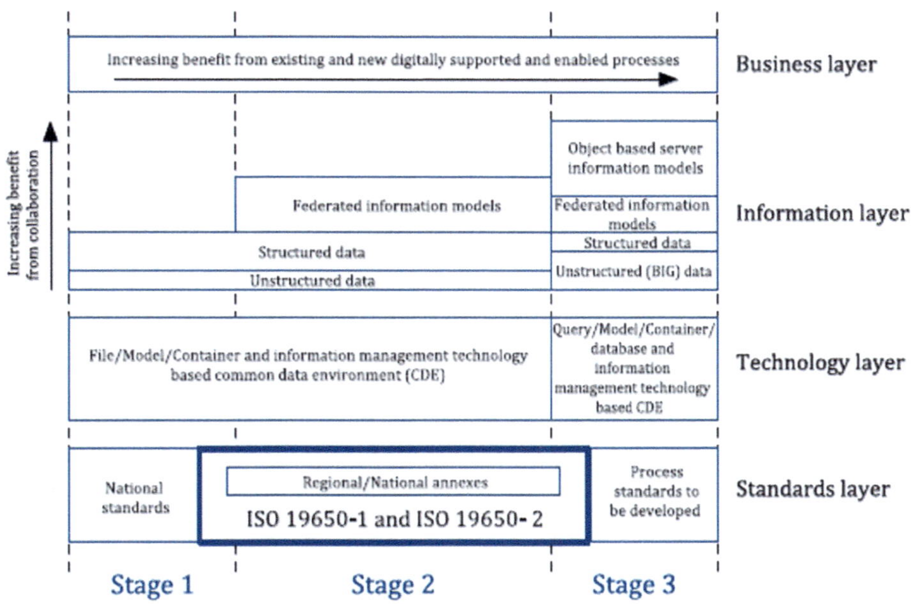

Figure 4.2 A perspective on stages of maturity of analogue and digital information management (Source: ISO 19650-2)

[Permission to reproduce extracts from British Standards is granted by BSI. British Standards can be obtained in PDF or hard copy formats from BSI Knowledge: https://knowledge.bsigroup.com or by contacting BSI Customer Services for hardcopies only: Tel: +44 (0)20 8996 9001, Email: cservices@bsigroup.com.]

4.2.2 BIM Level of Development (LOD)

LOD: The *LOD Specification* is a reference that enables practitioners in the AEC Industry to specify and articulate with a high level of clarity the content and reliability of BIMs at various stages in the design and construction process.

Level of Detail or Level of Model Detail: Level of detail is essentially how much detail is included in the model element. In other words, it is the description of graphical content of models at each of the stages defined for example in the Construction Industry Council (CIC) Scope of Services. To aid in understanding, the granularity of information in the foundation walls is presented in Figure 4.3.

Levels of Model Information (LOI) or Level of Information Detail: This is the description of non-graphical content of models at each of the stages defined for example in the CIC Scope of Services.

Level of Definition: This is a collective term used for and including 'level of model detail' and the 'level of information detail.'

The concept of level of development has been well-covered in PAS1192-2: 2013 with diagrammatic representation presented on Page 35 of the same document.

In the recent ISO 19650 standards, 'Level of Information Need' (no acronym) is a more generic term used in place of level of model definition/level of detail/level of information. The standard defines 'Level of Information Need' as a framework which defines the extent and granularity of information. Details about 'Level of Information Need' as stated in ISO 19650 can be further explored in BS EN 17412-1:2020.

4.2.3 nD Modelling

The production of building models is not the means to an end. A geometric model without data is not BIM, as this can already be done in existing CAD systems. In the context of BIM, attaching data on the geometric model is imperative. Some examples of data that can be attached to the geometric model include time and cost data. When time is added to the geometric model, it is called 4D-BIM. On the other hand, when cost is added, it is called 5D-BIM.

While there is a consensus of the meanings of 1-5D BIM dimensions in the industry, from n=>6, there is a lack of a common understanding of the different definitions. In some

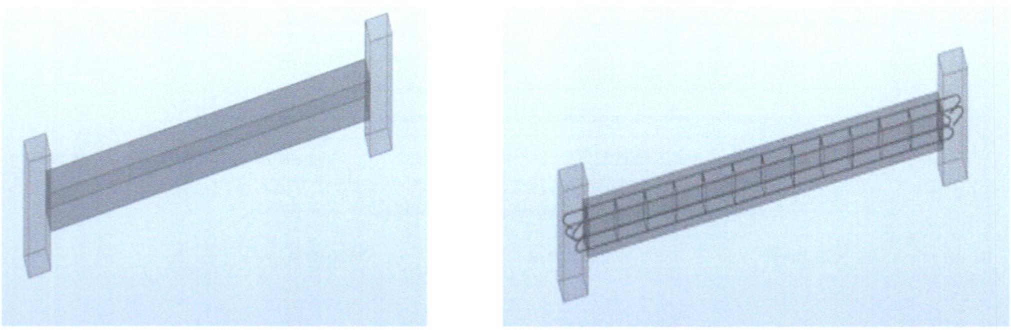

LOD-300 Beam
This model does not include reinforcement.

LOD-350 Beam
This object contains reinforcement.

Figure 4.3 Level of detail (Source: Author's)

publications, 6D may mean BIM for facilities management, while in others it may mean BIM for sustainability. In others, 7D may mean BIM for facilities management or BIM for sustainability. Table 4.1 depicts the different dimensions of BIM.

Generalizing the numbering representing 1, 2, ... in the form of dimensions, the concept of nD modelling is often used. The number n belongs to the set of integers.

In summary, the dimensions of BIM, denoted by 'n,' represent the different BIM applications in construction practice. It is important to note that there are endless applications in practice. The different common applications are cost, time, clash detection, sustainability, code compliance, health and safety, construction waste, traffic control, fire, etc. Each application is underpinned by its type of data that is required to be embedded in the 3D model. It is important to note that the dimensions may not necessarily be cumulative. For example, adding facilities management data (for 6D or 7D) does not require time (4D) and cost (5D) data to have been added to the model. It could be a simple straightforward addition of facilities management data on a 3D model. In fact, in other to obtain an accurate 4D BIM, it is imperative to first determine the quantities of construction materials which in conjunction with rates, the duration of activities is determined. The activities time or schedule is then attached to the geometric model to form 4D-BIM. Determining quantities, often known as quantity take-offs is constitute part of 5D modelling.

4.2.4 BIM Team

The recent ISO 19650 defines the various BIM team members as follows:

Appointing Party: The organization leading the project or asset management. For a project, this is typically the client, who may also be the asset owner.

Lead Appointed Party: The party who is accountable for coordinating information exchange between task teams or between a delivery team and the appointing party.

Appointed Party: Anyone generating information about the project, for example, a contractor, sub-contractor, supplier, and consultant.

Project Team: Everyone involved in the project, regardless of appointment/contract arrangement.

Delivery Team: A lead appointed party and their task teams, for example, a contractor and its sub-contractors and suppliers.

Task Team: A person or group of people performing a specific task, for example, the architecture team or sub-contractor who is designing/constructing curtain walling.

4.2.5 BIM Periodic Table

The BIM periodic table takes its inspiration from the chemical periodic table. Its goal is to facilitate the understanding of concepts of BIM. The table contains all the main elements of BIM in an easy-to-follow, visual reference – ideal for printing or sharing (as a .pdf) or displaying on a big screen (via our interactive).

The main BIM elements in the BIM periodic table have been grouped into Strategy, Foundations, Collaboration, Process, People, Technology, Standards, Enable Tools, and Resources categories.

Strategy: In this component, a company interested in adopted BIM should set a vision and define a strategy of how to attain the vision. In order to attain the vision, it is imperative to

Table 4.1 BIM dimensions

Dimensions	*Pictorial representation*	*Descriptions and examples*
1D	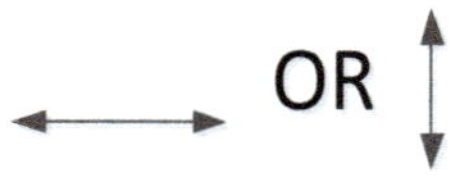	1D refers to objects or entities that exist only along a single dimension. In other words, it has only length and no other dimension(s).
2D	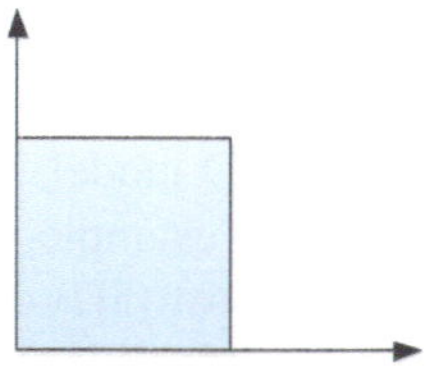	2D refers to objects or entities that exist within a two-dimensional plane. Contrary to 3D objects that have length, width, and height, 2D objects have only length and width, meaning they are flat and lack depth. A square and a rectangle are examples of 2D objects.
3D	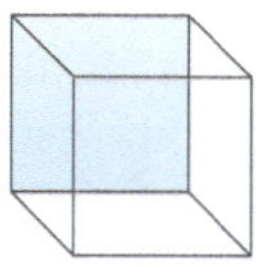	3D refers to objects or entities that exist in three spatial dimensions: length, width, and height or dept. A cube is an example of a 3D object.
4D	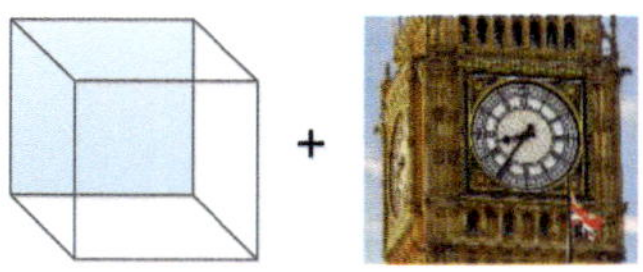	Scheduling
5D	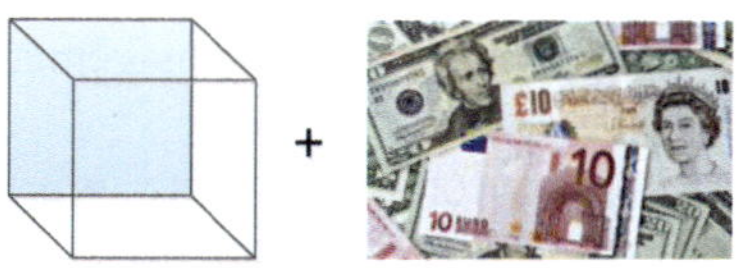	Cost
6D		Facilities management; Data types (installation date, replacement date)
7D		Sustainability; Data types (embodied energy and CO_2 intensities)
8D		Health & Safety

consider supporting foundations, processes, technology, tools, and people that you require that will inform how and when to implement the strategy.

Foundations: This is about laying a robust foundation of efficient systems for communication, information exchange, and data transfer, to underpin advanced BIM processes.

Collaboration: This component is about establishing better and more efficient ways of working with a focus on collaboration. It is important to consider digital tools that can aid in effective collaboration.

Process: This is about identifying where and how to improve in your current processes.

People: This element deals with the provision of clear communication to colleagues as to why and how you intend to implement BIM and gain support by engaging BIM champions, as well as getting support from senior management.

Technology: This element is about making sure, a company about to implement BIM must make it has the right technology, software, and hardware, to support your BIM aims and objective

Standards: This element is about considering the standards, procedures, and supplementary documents available that will assist with your strategy and help achieve collaborative BIM.

Enable Tools: This element is about considering enabling tools that can help design, develop, deliver, and maintain the built asset. In doing this, it is important to consider free tools to minimize cost and only go for costly ones when the former is not available.

Resources: This element strongly encourages firms implementing BIM to explore information by considering the available resources. Some of the resources can be found on the internet and social media. The internet and social media have created a valuable online community of support.

4.3 System Integration

Having reviewed the various common technologies used in construction, it is imperative to examine two main concepts that allow for information exchange or sharing between them. These are integration and interoperability. Managing information during the lifecycle of a construction project is becoming more and more important in contemporary construction project management with an increasing amount of information being produced by AEC. This information comprises of building geometry, spatial relationships, and quantities and properties of building components. Efficient processing of these kinds of information to support different AEC professionals is essential for a successful project management. To enhance understanding, the terms 'component' and 'system' relevant to these two concepts will have to be explained. According to IEEE 610, components are one of the parts that make up a system, while a system is a collection of components organized to accomplish a specific function or a set of functions.

The term integration refers to the process of combining components into an overall system (after IEEE 610). In other words, system integration refers to the process of connecting two or more applications so that data can be shared between them, allowing information to be read or edited on any connected device. This is achieved using connectors or middleware, which act as intermediaries to facilitate the transfer of data between different systems. In the context of digital businesses, system integration is crucial for enabling the seamless flow of information, fostering effective collaboration, and coordinating communication across various channels. The primary goal of system integration is to create a cohesive system where data can be shared widely, irrespective of the technologies used in sending, receiving, and distributing that information. Five different BIM integration examples have been examined in Wastiels and

Decuypere (2019). The first is about how different building components (with their quantities) are manually linked to predefined LCA profiles available in the LCA software database or creates new LCA profiles where needed. The second method is about how BIM data are imported data (in IFC format), and linked to predefined LCA profiles available in the LCA software databases. Thirdly, the BIM Viewer is used ion linking LCA profiles. Fourthly, LCA plugin is used to link LCA profiles to BIM objects within the native BIM environment. Lastly, LCA information is included in the BIM objects that are used in the BIM model.

4.4 System Interoperability

According to IEEE 610, interoperability is the ability of two or more systems (or components) to exchange and subsequently use their information. Consequently, interoperability is concerned with the ability of systems to communicate – and it requires that the communicated information can be understood by the receiving system – but it is not concerned with whether the communicating systems do anything sensible as a whole. The interoperability between two systems could be fine, but whether the two systems as a whole actually performed any useful function would be irrelevant as far as the interoperability is concerned.

Syntactic interoperability refers to the ability of two (or more) separate systems or software programmes to communicate and exchange data (or information) with each other and use the data (or information) that has been exchanged (Bahar et al., 2013; Rezaei et al., 2014). The concept of 'Interoperability' in the broader BIM context refers to the seamless exchange of BIM data between different software applications, which is facilitated by using industry-standard file formats:

- **Industry Foundation Classes (IFC):** IFC is a standardized, digital description of the built asset industry. It is an open, international standard (ISO 16739-1:2018) and promotes vendor-neutral, or agnostic, and usable capabilities across a wide range of hardware devices, software platforms, and interfaces for many different use cases.
- **Green Building eXtensible Markup Language (gbXML):** It is an industry-supported schema for sharing building information between desperate building design software tools.
- **Construction Operations Building Information Exchange (COBie):** COBie is an information exchange specification for the lifecycle capture and delivery of information needed by facility managers. COBie can be viewed in design, construction, and maintenance software as well as in simple spreadsheets.

  Collaboration on projects strongly depends on interoperability of software systems. In a BIM-compliant project, data can be exchanged using different formats. For example, the contractor can exchange information with a BIM platform using IFC. To be precise, an IFC model can be loaded to Autodesk Construction Cloud and a contractor can accessed the model on site using a smart device.

Efficient processing of information depends largely on how the computer programmes interface (i.e., integration) and/or how easy such information can easily be exchanged between two or more computer systems (i.e., interoperability).

The importance of system integration and interoperability has led to a surge in research in the same of late. Valero and Adán (2016) focused on the integration of BIM-related technologies such as RFID, 3D laser scanning, cameras, GIS, and GPS. The authors went on to state that 'there is a general lack of studies on interoperability of data among different BIM software programmes and hard data-acquisition technologies' (Chen et al., 2015). Most recent papers concerned with BIM technology integration focus on specific issues such as sustainability (Cheng et al., 2024) or more mainstream topics such as GIS (Cao et al., 2023), or sensor technology (Hu et al., 2024). Also, emphases have been placed on new technologies such as AR/VR (Monla et al., 2024), robotics (Zhu et al., 2024), and the IoT (Chen et al., 2023) and how they can be successfully integrated with BIM to improve the management of construction information in the industry. Taking the virtual spectrum further, integrating robotics into BIM could represent the next step towards the correct design and application of automated construction processes in the AEC. Despite recent increase in research about BIM integration with other technologies, the emerging nature and multiplicity of ICT means further research is still needed. The main questions: How does BIM interfaced with other ICTs (e.g., robots, 3D laser scanners, 3D printers, etc.)? Do professionals reflect on the format of their output data during the processing of the same? Do professionals consider other phases in the project lifecycle during the processing of their data?

Interoperability in BIM can be likened to the concept of the Tower of Babel story. In Genesis 11:1–9 of the Biblical Book, a myth is used to explain why the peoples of the world speak different languages. In the verses, it is argued that the people of Babel once spoke a common language and agreed to build a tower tall enough to reach God in Heaven.

By speaking a common language, the community worked in harmony and successfully coordinated the construction of the tower. However, it is said that, at some point, God became displeased and introduced confusion among them by causing them to speak different languages, making it impossible for them to understand one another. This disruption ultimately halted their progress, preventing them from achieving their ambitious goal of constructing a tower tall enough to reach God.

In the narrative, the community's shared language allowed for seamless communication and collaboration, enabling them to work towards a unified goal with great efficiency. Similarly, in BIM, interoperability ensures that various tools, software, and stakeholders involved in a construction project 'speak a common language,' facilitating the smooth exchange and understanding of information across different platforms and disciplines.

However, just as the introduction of different languages among the builders caused confusion, lack of interoperability in BIM leads to communication barriers, inefficiencies, and errors. Different software systems or file formats that cannot effectively exchange or interpret data create 'language barriers' within a project team. This disrupts collaboration, delays progress, and jeopardizes the achievement of project goals.

Therefore, achieving interoperability in BIM is critical. It ensures that information flows seamlessly between systems, enabling all parties to work together efficiently towards a shared vision, much like the builders could before the confusion of languages. Interoperability underpins the success of BIM by fostering integration, reducing errors, and supporting the collaborative efforts required for complex construction projects.

4.5 OpenBIM

According to buildingSmart, openBIM is a collaborative process that is vendor neutral. OpenBIM processes can be defined as sharable project information that supports seamless collaboration for all project participants. OpenBIM facilitates interoperability to benefit projects and assets throughout their lifecycle. Some examples of openBIM standards include: IFC, BCF, and information delivery specification (IDS).

4.6 Conclusions

The exploration of 'Common Concepts, System Integration & Interoperability' in Chapter 4 brings to light the critical facets of digital transformation within the construction industry, emphasizing the importance of BIM technologies. This chapter has dissected the layers of BIM's maturity models, development levels, and the expansive realm of nD modelling, showcasing the progressive steps necessary for organizations to fully embrace and implement digital construction methodologies effectively. The insights into system integration and interoperability underscore the necessity for open standards and seamless communication across diverse technological platforms, ensuring that the data flow within a construction project is both efficient and coherent.

As we pivot towards the next chapter, 'Digital Construction Applications,' we aim to extend the theoretical framework established thus far by examining the practical applications of digital construction technologies. This forthcoming chapter promises to shed light on the myriad ways these technologies are being utilized in real-world construction projects. Emphasizing process maps and software implementation, the chapter seeks to demystify the principles and processes underlying digital construction applications. This approach not only contrasts with conventional research methodologies by providing a clearer, more accessible understanding of digital construction technologies but also emphasizes the practical benefits of these technologies in enhancing information exchange among project stakeholders. Through detailed process modelling and software demonstrations, readers will gain a comprehensive understanding of the interoperability of digital construction technologies and their potent capabilities in transforming project management, execution, and collaboration.

The upcoming chapter delves into the practical sphere, exploring the diverse applications of digital construction technologies. By presenting process maps and real-life software implementations, this chapter aims to bridge the gap between theoretical knowledge and practical application. It will provide a detailed examination of how digital tools and methodologies are applied across various stages of the construction process, highlighting the benefits of improved efficiency, accuracy, and communication. This exploration is designed to equip readers with a deeper understanding of the operational synergies possible through digital construction technologies, setting the stage for innovation and excellence in the construction industry.

Chapter 5

Digital Construction Applications

5.1 Introduction

In this chapter, the practical applications of digital construction technologies are explored in depth. This analysis serves as the foundation for understanding how these technologies interoperate and their effectiveness in enhancing information exchange among project stakeholders. The chapter takes a distinctive approach, setting it apart from much of the existing research in two key ways.

First, the application steps are modelled using process maps, providing readers with a clear visualization of the principles and workflows involved. Second, these processes are implemented within software environments, simplifying their interpretation and practical application.

To ensure clarity, the chapter is structured into two main sections. Given the pivotal role of building information modelling (BIM) in driving construction's digital transformation, Section 5.2 focuses extensively on this topic. Section 5.3 delves into the integration of BIM with other digital technologies to optimize information productivity.

5.2 Applications of BIM in Practice

5.2.1 *BIM Design*

5.2.1.1 Design or Development Models from Scratch

The main activity in BIM design is to develop a model that meets client's requirements and that will be executed on-site. The main activities during design include conceptualization, designing, and visualization. The components used in design could be designed from scratch or imported from BIM object libraries[1234]. The BIM object libraries are populated by manufacturers with their products that can used in design projects. Once a design is accepted based on the products used from the library, the products are easily procured directly by a developer or contractor and used in the delivery of the project. In fact, manufacturers are producing digital replicas of their products, which can then be used in designing BIM models of real-life projects. Designing using any BIM authoring tools is akin to developing a house using Lego blocks in both approaches.

Figure 5.1 illustrates the first approach where the design is all within a BIM authoring tool, in this case Revit.

In this case, the components such as walls, doors, and windows are obtained from the system. In case some of the components are not available, then it is obtained from the family library that comes with Revit installation. It is important to note that in earlier Revit versions up to

DOI: 10.1201/9781003476740-5

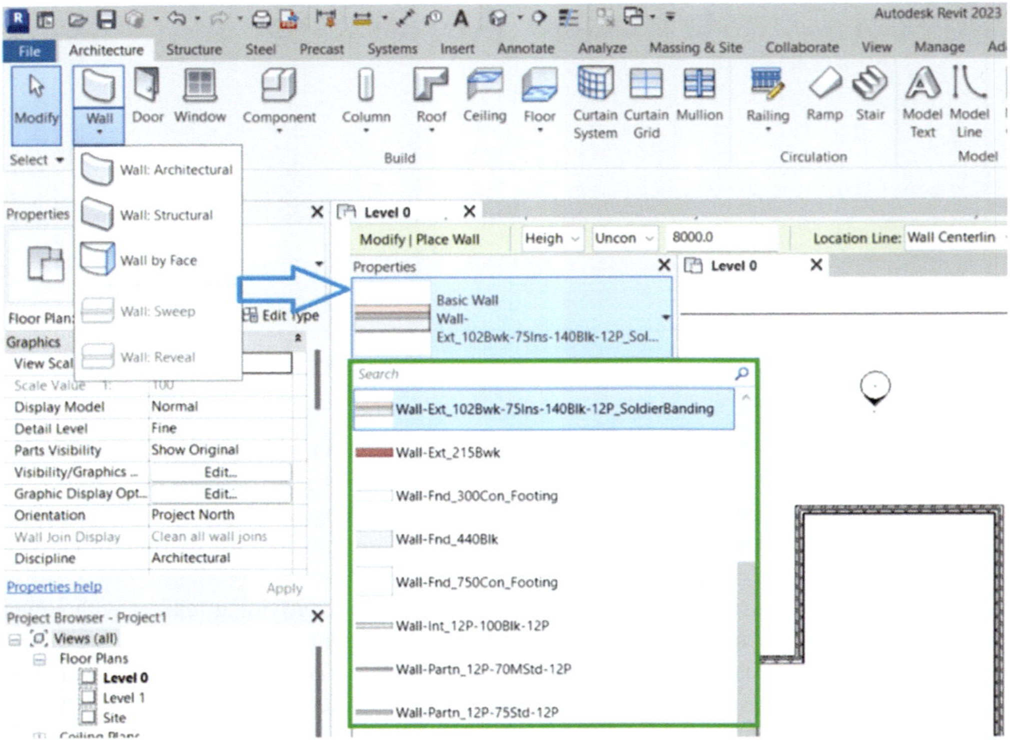

Figure 5.1 Design in Revit environment using system objects

and including 2021, the Revit family library was often installed locally on personal computers (PCs). These components were accessible through 'Load Family' application in previous Revit versions. From 2022, the components were moved to an online platform but still linked directly to the Revit software. The components can be loaded through the 'Load Autodesk Family' application in Revit 2022. When this operation is launched, it takes users online to the online Autodesk family platforms where components are then pulled and used in developing BIM models (Figure 5.2).

In the second option, components are brought from an external online platform not linked to the Revit software such as 'Load Autodesk Family.' Some examples of these platforms include: Revit City, BIM Object, and NBS libraries. The advantage of pulling components from these sources is that it reflects the kind of components that are in the market. Should the design be agreed upon by the client, then it will be easy to procure the materials or components for the construction of the building.

In Figure 5.3, designing using objects from the NBS library is presented. The object is imported through the use of a plug-in developed by NBS.

5.2.1.2 Design or Development of Models from Existing Data

The three common data sources or existing models from which BIM models can be developed are from PDF, DWG, and point cloud data.

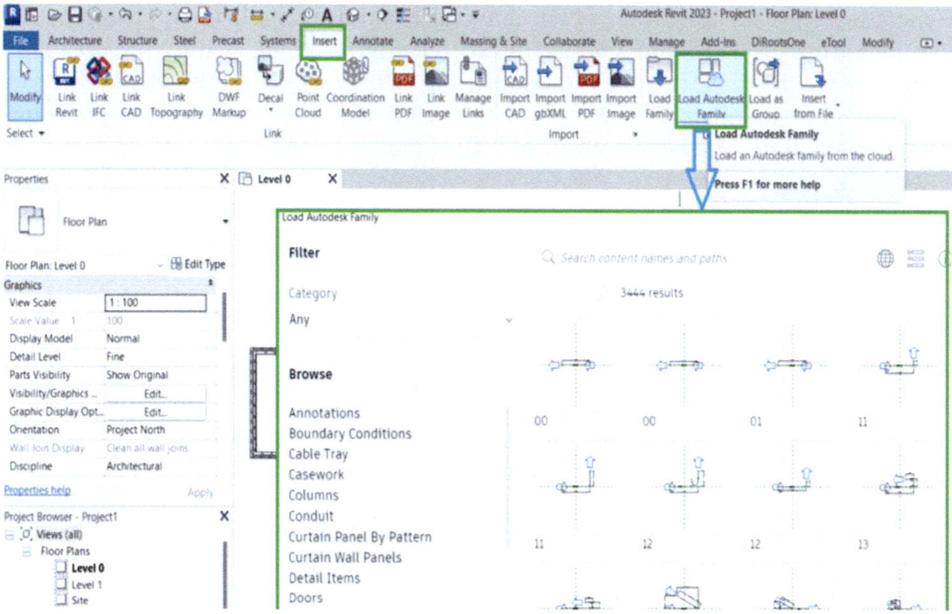

Figure 5.2 Loading components from the Autodesk Revit family Library

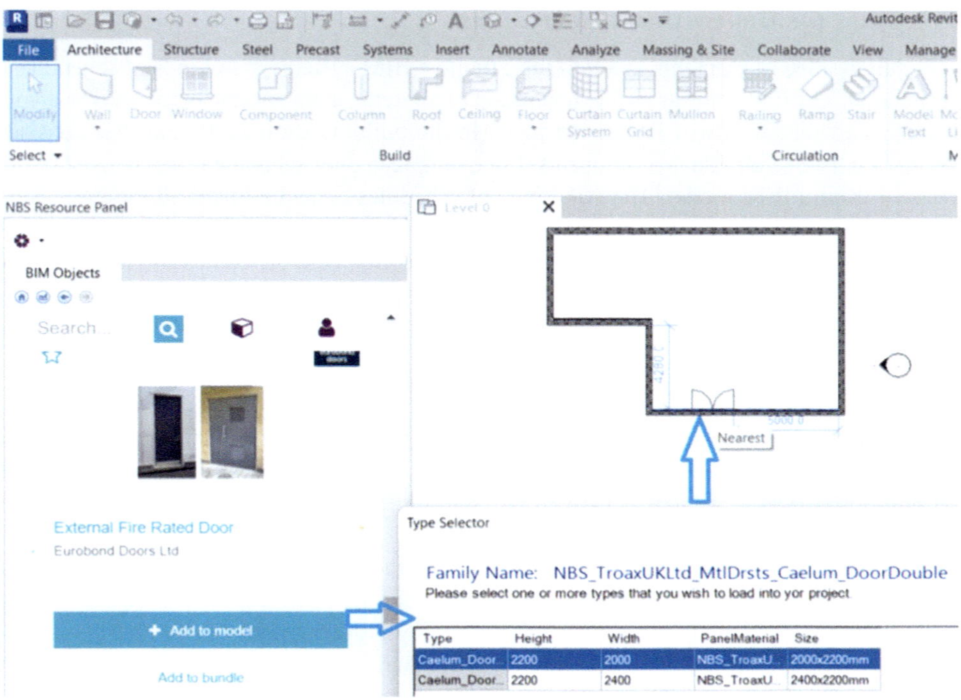

Figure 5.3 Design in Revit environment using an Object Library

Generating BIM models from a PDF drawing entail retracing over it. The process include: importing the PDF through the 'Insert' function in Revit, scaling the drawing, retracing over the drawing, removing the importing PDF and leaving the 3D BIM model in Revit. It is important to note that the process for producing BIM models from PDFs are the same for image files. The development of a BIM model from a PDF drawing is presented in Figure 5.4.

The Ground Floor on Figure 5.4 is then retraced and appropriate components, e.g., doors are added. Once the designer is satisfied, the PDF is removed using the Manage Links function of the Insert tab.

Using the same approach as in the case of PDF drawing a DWG file can be imported or linked into Revit as shown in Figure 5.5.

The DWG drawing in Figure 5.5 can now be retraced, and once the expected model is obtained, the DWG file can be removed through the Manage Links function in the Insert tab.

The point cloud model compatible file format with Revit is ReCap (RCP). The model in RCP format from Autodesk ReCap has been used in this example, imported using the same approach for PDF and DWG (see Figure 5.6).

The point cloud (i.e., RCP format) in Figure 5.6 can now be retraced to develop the expected BIM model. To ensure an accurate or near-accurate tracing, use the View Range function to adjust and improve visibility as indicated in Figure 5.7.

Once the expected model is obtained, the RCP point cloud can be removed through the Manage Links function in the Insert tab (see Figure 5.8).

5.2.2 BIM for Generation of Schedules

A schedule is a list of elements in a building. In most BIM systems, two kinds of schedules can be generated. The first consists of the generation of all items in the model, often known as multi-category schedule. This will list all the items in the building project. The second consists of generating a list of individual components. For example, a door schedule refers to a list of doors generated from a BIM model. The two kinds of schedules generated from Revit are presented in Figures 5.9 and 5.10.

It is also possible to generate multi-category and component schedules in iTwoCostX [see Figures 5.11 and 5.12].

5.2.3 BIM for Quantity Take-Off (QTO)

QTOs are a detailed measurement of materials and labour needed to complete a construction project. The units of these measurements can be m, m^2, m^3, etc. QTO can be conducted in BIM authoring tools (e.g., Revit) as well as in application software such as Navisworks and iTwoCostX.

Like schedules in Revit, QTO can be generated for the whole building known as multi-category QTO or component-based QTO. As an example of the later, a QTO can be conducted on any element, e.g., walls only. A wall QTO is shown in Figure 5.13.

In Navisworks, QTO can be conducted on individual components as well as on the whole building. At the component level, QTO can be generated and inserted in work break-down elements of standard measurement methods such as Uniformat. The process is indicted in Figure 5.14.

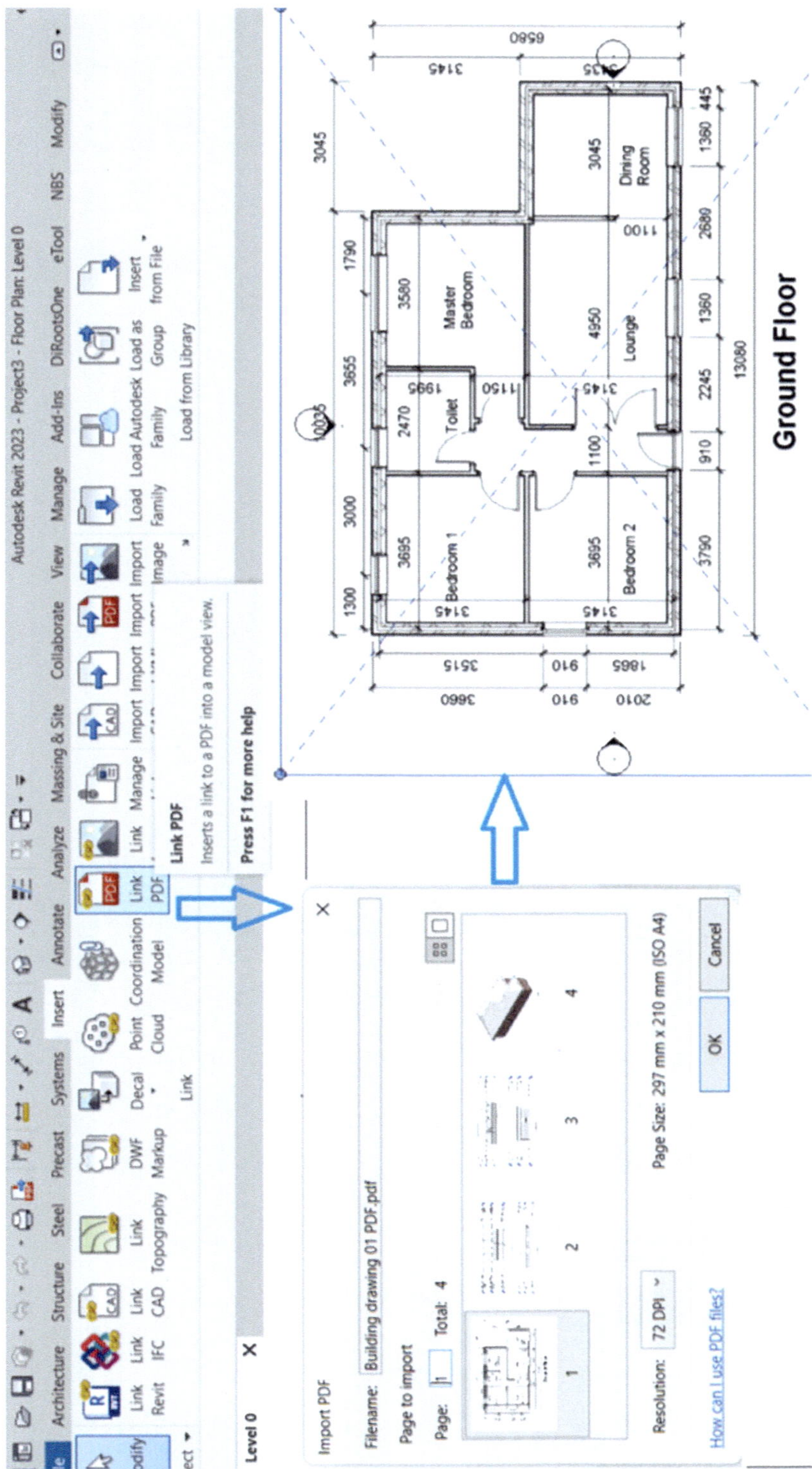

Figure 5.4 Modelling from a PDF a drawing

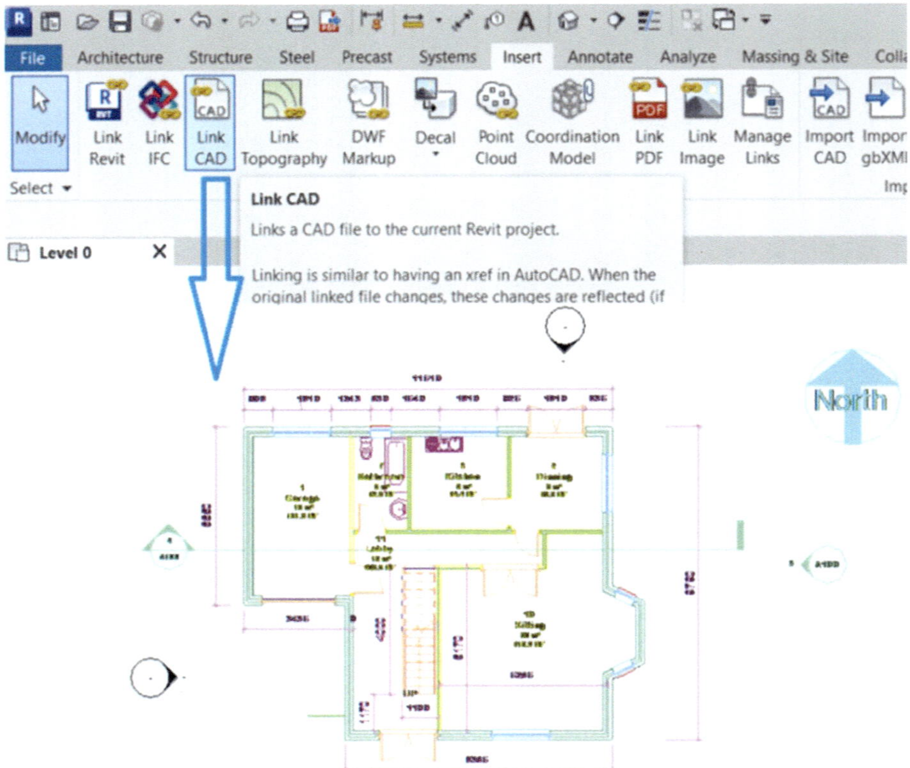

Figure 5.5 Modelling from a .DWG drawing

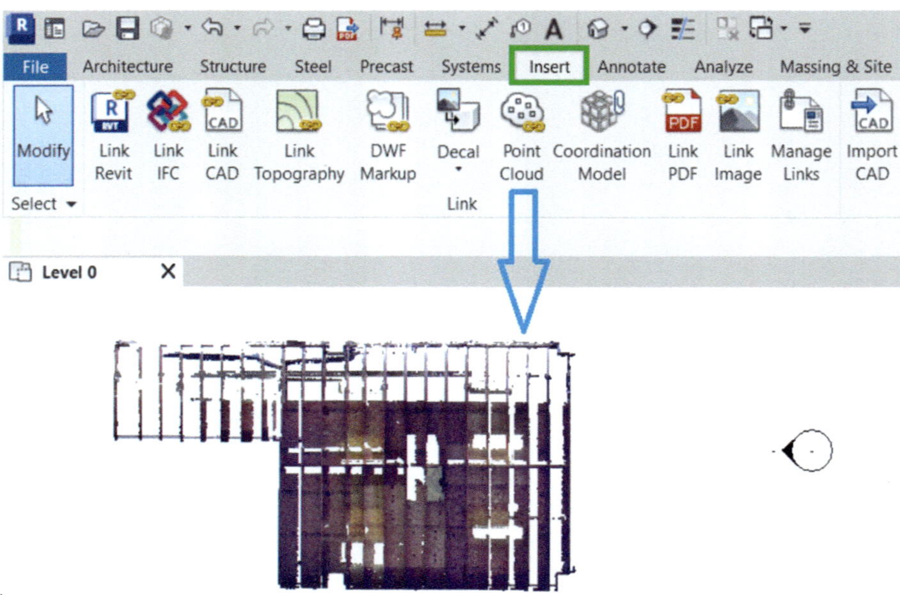

Figure 5.6 Modelling from point cloud data

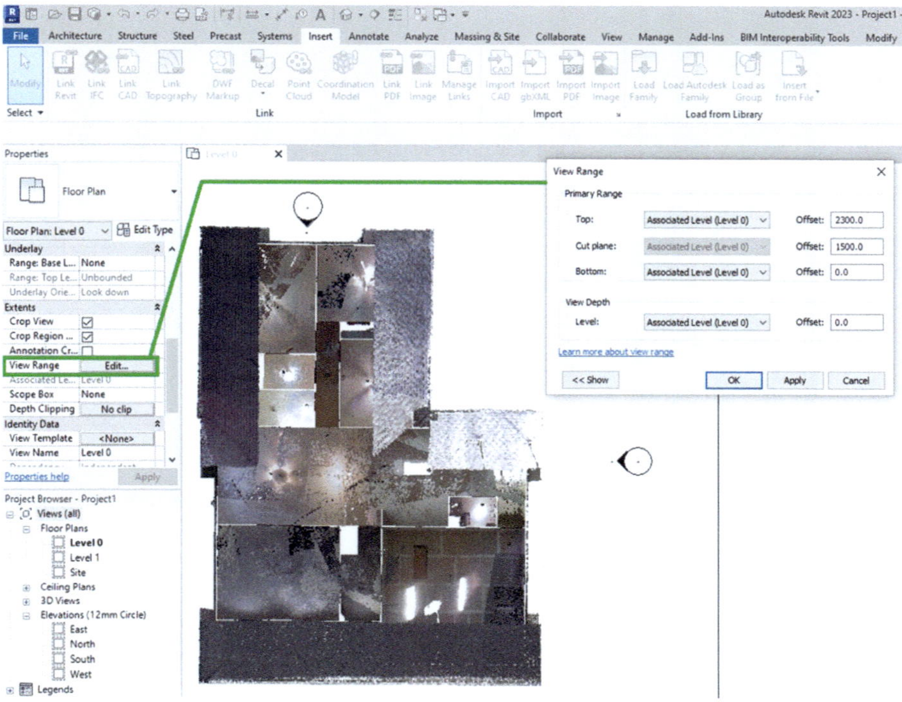

Figure 5.7 Improving visibility of the point cloud data

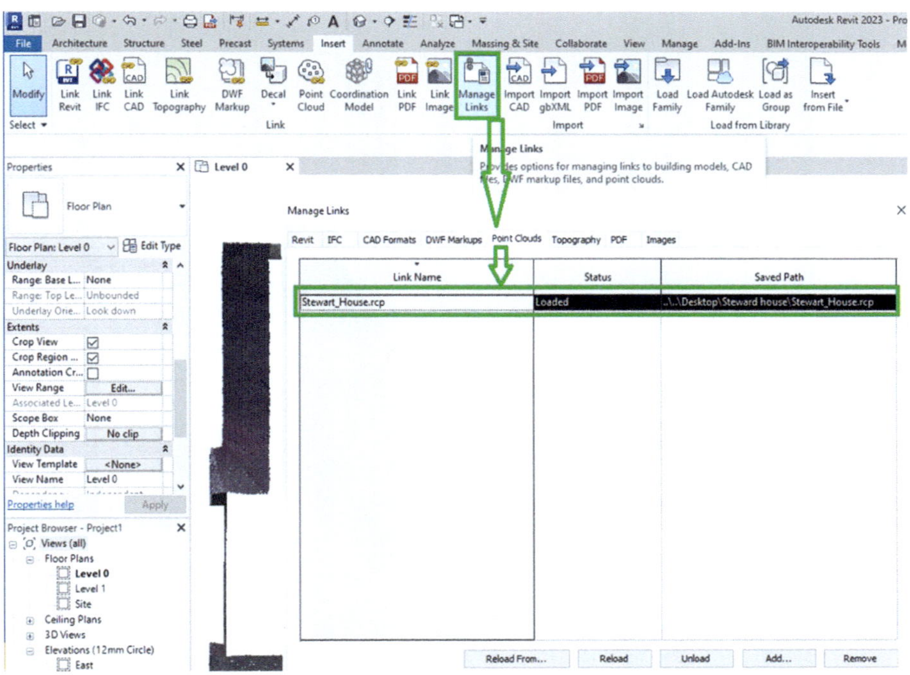

Figure 5.8 Removing the point cloud from the background

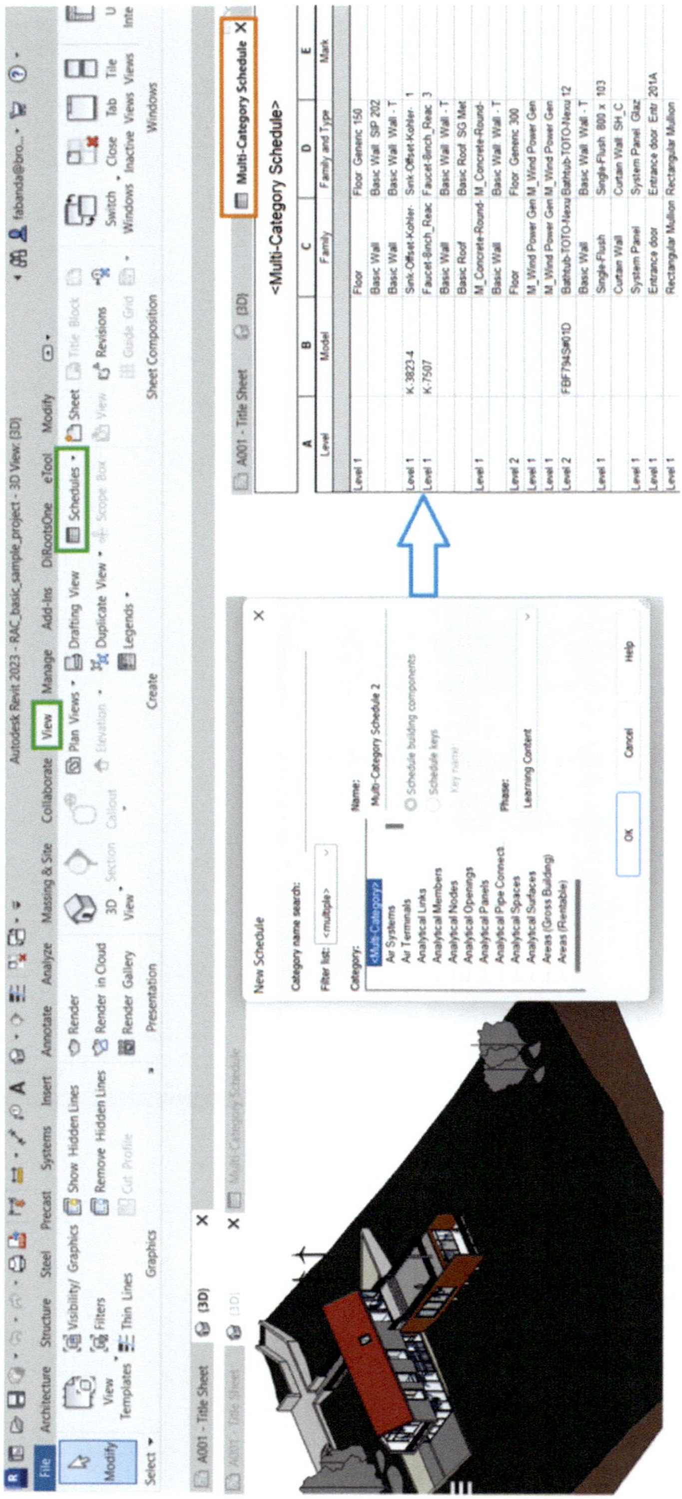

Figure 5.9 Multi-category schedule [in Revit]

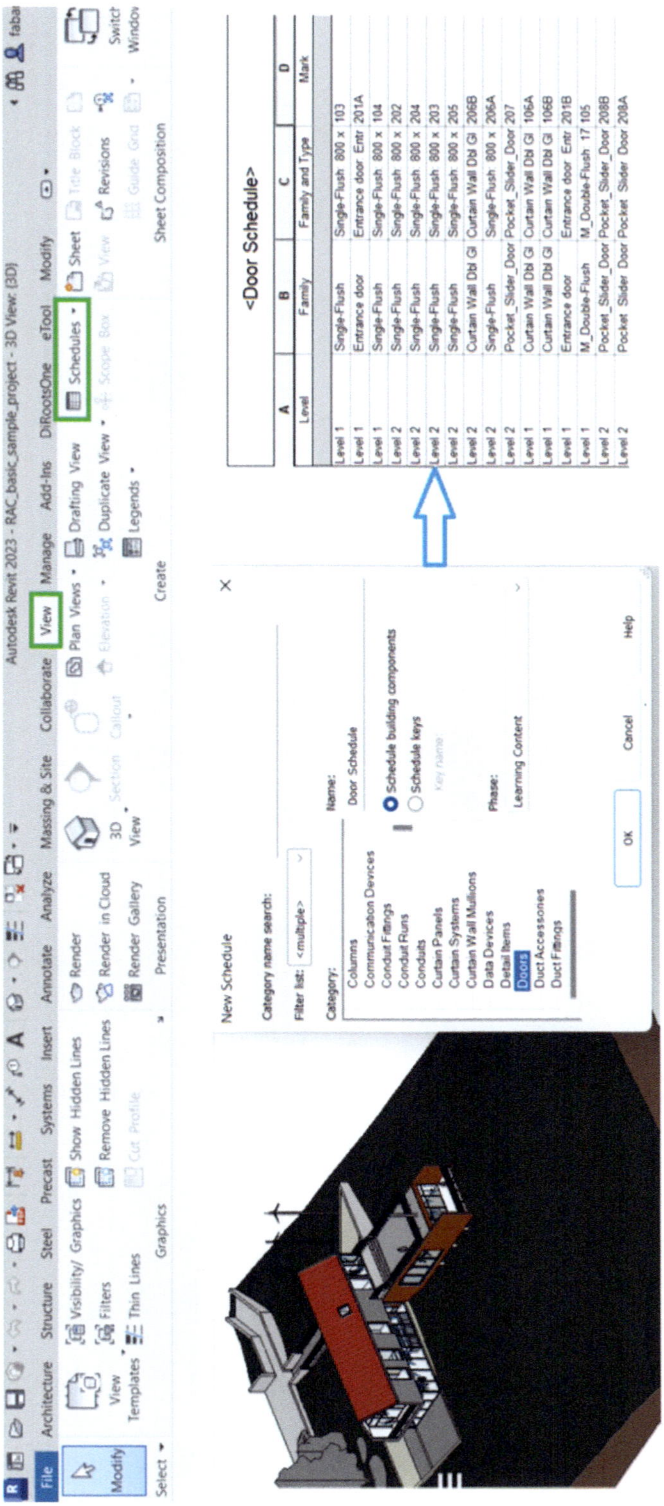

Figure 5.10 Component schedule [in Revit]

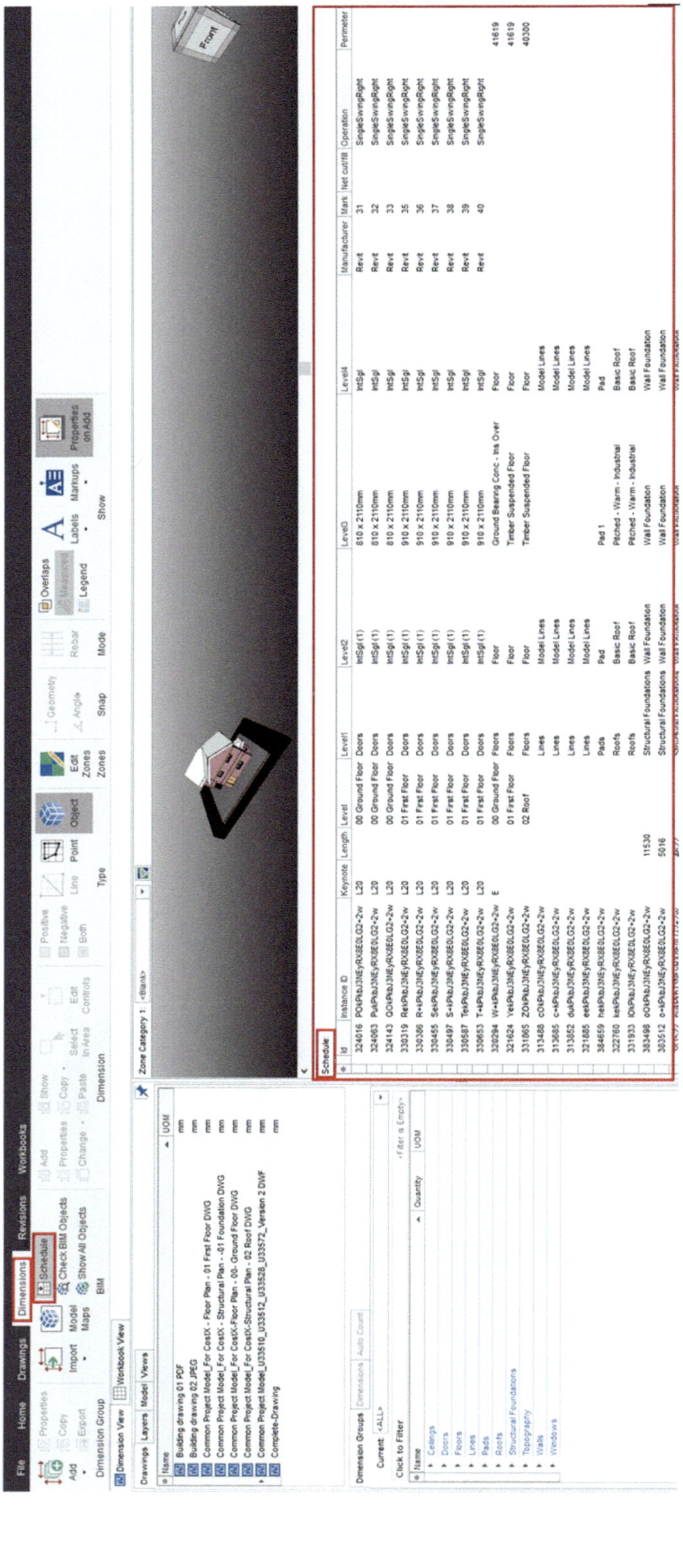

Figure 5.11 Multi-category schedule [in iTwoCostX]

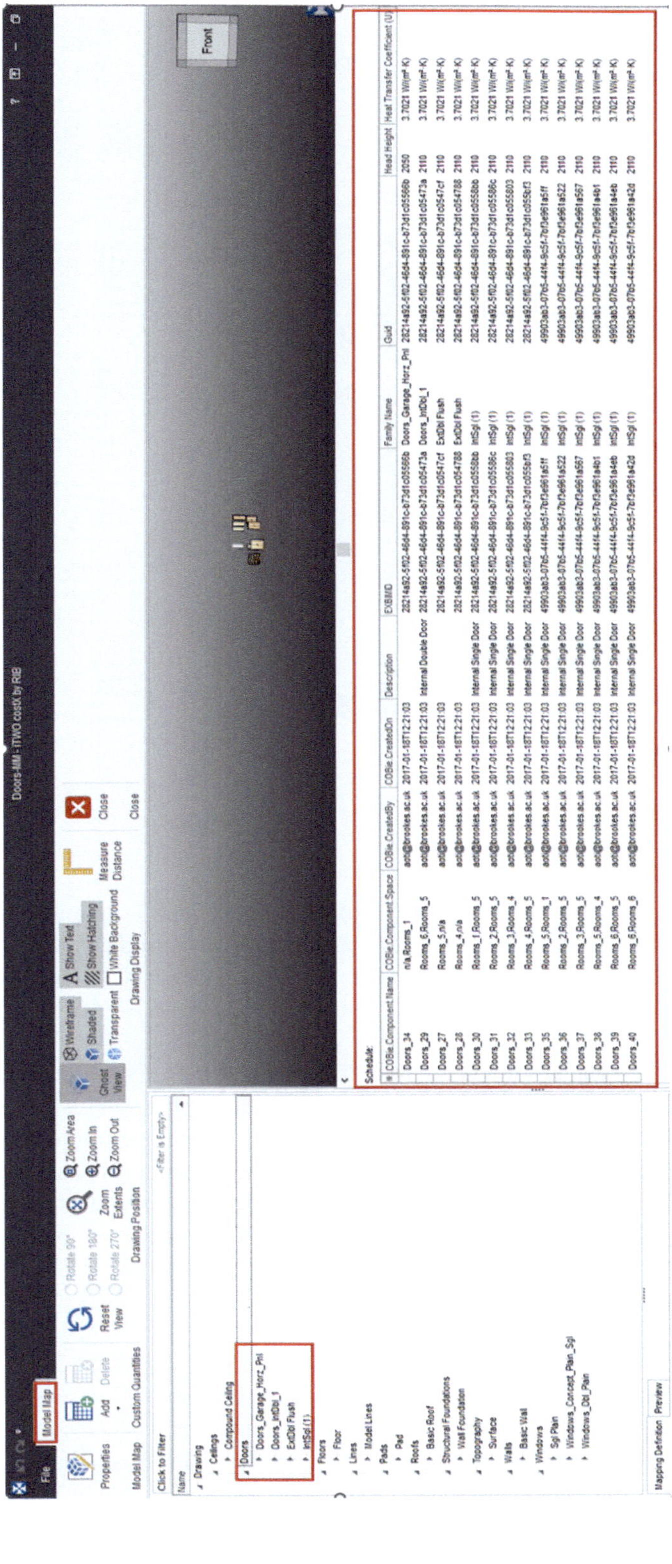

Figure 5.12 Component schedule [in iTwoCostX]

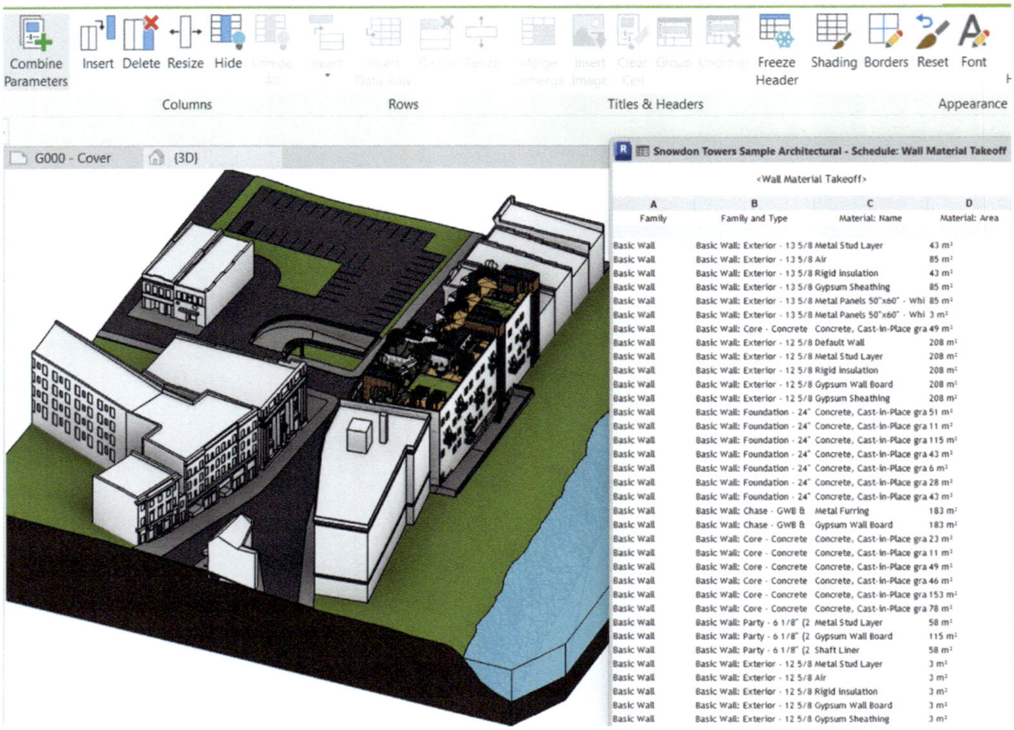

Figure 5.13 Walls QTO in Revit

In the second method, the QTO from the building is generated in one goal and displayed. In this case, it is not possible to insert the generated data into a specified WBS in the standard measurement methods. The generated QTO can then be exported into MS Excel that can be used for other applications. The process for multi-category QTO in Navisworks is presented in Figure 5.15.

In iTwoCostX, QTOs can be computed from PDF or image files, CAD files and BIM models. The different QTO techniques will be examined in the ensuing paragraphs.

QTO from PDF: In PDF, it is always important to check the scale of the drawing and then calibrate it. This is because the drawing might have been squashed and the scale distorted.

QTO from CAD: In CAD, there is no need to calibrate as the design from CAD is designed on a 1:1 scale. The layers in CAD also present great advantages for exploring the model.

QTO from BIM: In iTwoCostX, QTO can be conducted in 3 ways.

Method 1: Using Model Maps

In this method, specific data are extracted from a building component in iTwoCostX. Upon extraction, it is simply previewed in a preview window and not necessarily inserted in the Dimension Group. It is a method simply aimed at determining the quantity of a building component, with no goal of inserting in the Dimension Group and/or Workbook (see Figure 5.16).

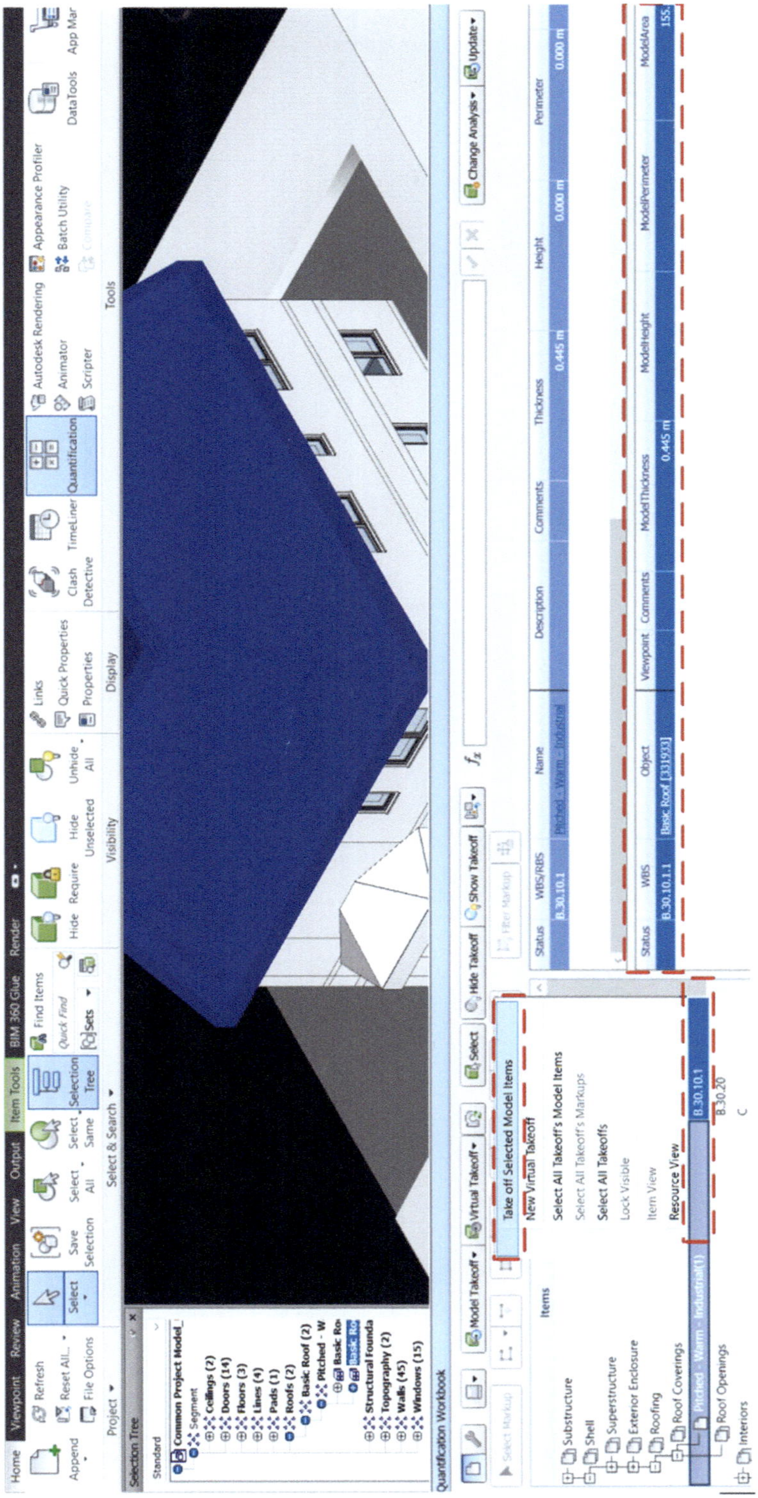

Figure 5.14 Component QTO in Navisworks

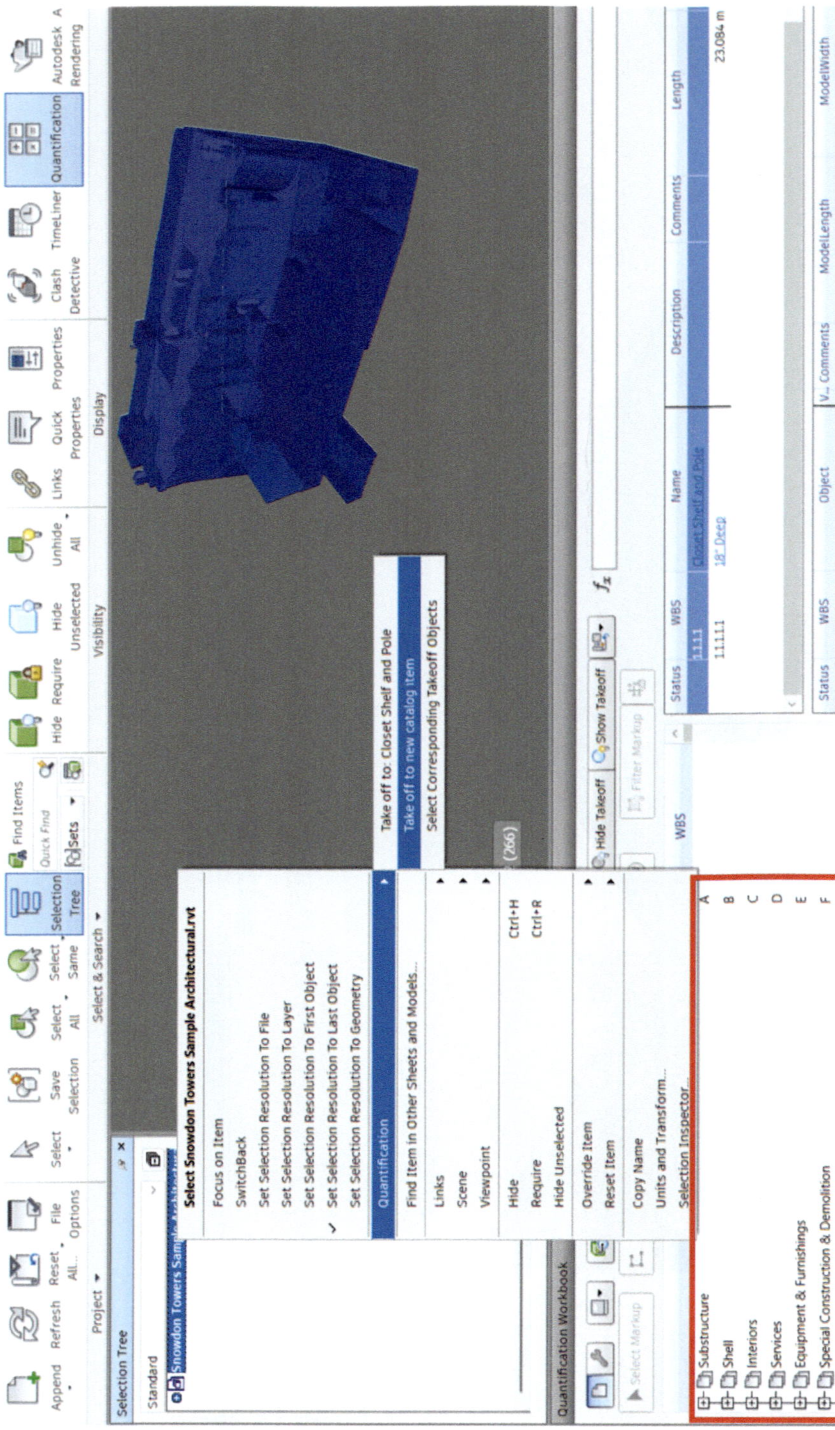

Figure 5.15 Multi-category QTO in Navisworks

Figure 5.16 QTO in iTwoCostX-Model Maps

Method 2: Automatic computation of QTO from specific components in iTwoCostX

This method is used to compute the quantities of a given component of a building in iTwo-CostX. The method is very useful in determining quantities from objects with very irregular shapes. Although there are many steps in implementing this method, the most important step is the creation of the quantity dimension group from the object. This is achieved by right clicking as shown in Figure 5.17, instead of the conventional approach from the Dimension icon.

Method 3: Using model template

In this method, a model template is used and the quantities are automatically extracted from the whole model in a very few steps (see Figure 5.18). The template comes with the installation of iTwoCostX.

5.2.4 BIM for Cost Estimation

The main tasks in quantity surveying are quantity take-offs, cost estimating, and bidding for projects.

The process for cost estimation involves developing a BIM model, exporting into a 5D BIM software, e.g., iTwoCostX and then developing cost plans. The process of developing cost plans in Revit and iTwoCostX is presented in Figure 5.19.

The generalized 5B BIM workflow is presented in Figure 5.20.

5.2.5 BIM for Clash Detection

In design terms, a clash occurs when components that make up a built asset are not spatially coordinated and conflict. In a BIM process, these clashes can be spotted more easily during the design phase of a project ahead of work starting on site.

Clashes arise as the models of various disciplines – architectural, structural, mechanical, electrical, plumbing, fire-fighting, etc., are combined to make a composite model. This is often due to the fact that models might have been developed by different consultants using the same or different computers. Finding the clashes before starting the project is important as it has been proven too costly to detect clashes on-site and fix or resolve them.

Hard Clash: This type of clash occurs when objects actually touch or intersects. For example, a column running through a wall or pipework through a steel beam.

Hard Conservative Hard (Conservative). This option performs the same clash test as Hard; however, it additionally applies a conservative intersection method (Autodesk, 2018).

Intersection Method: A standard Hard clash test type applies a normal intersection method, which sets the clash test to check for intersections between any of the triangles defining the two items being tested (remember all Autodesk Navisworks geometry is composed of triangles). This may miss clashes between items where none of the triangles intersect. For example, two pipes that are exactly parallel and overlap each other slightly at their ends. The pipes intersect, yet none of the triangles that define their geometry do and so this clash would be missed using the standard Hard clash test type. However, choosing Hard (Conservative) reports all pairs of items, which *might* clash. This may give false positives in the results, but it is a more thorough and safer clash detection method (Autodesk, 2018).

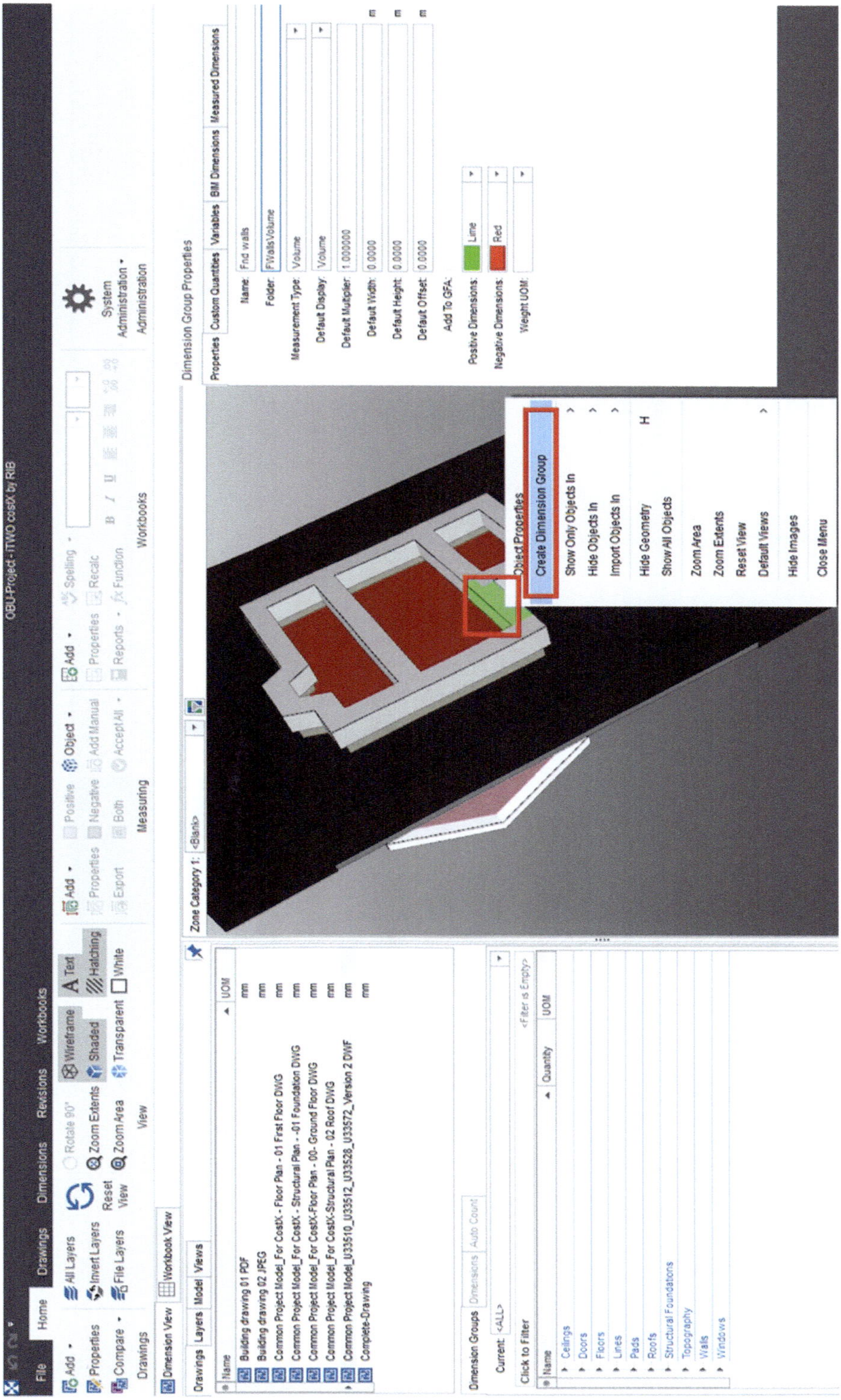

Figure 5.17 Component extraction from an object

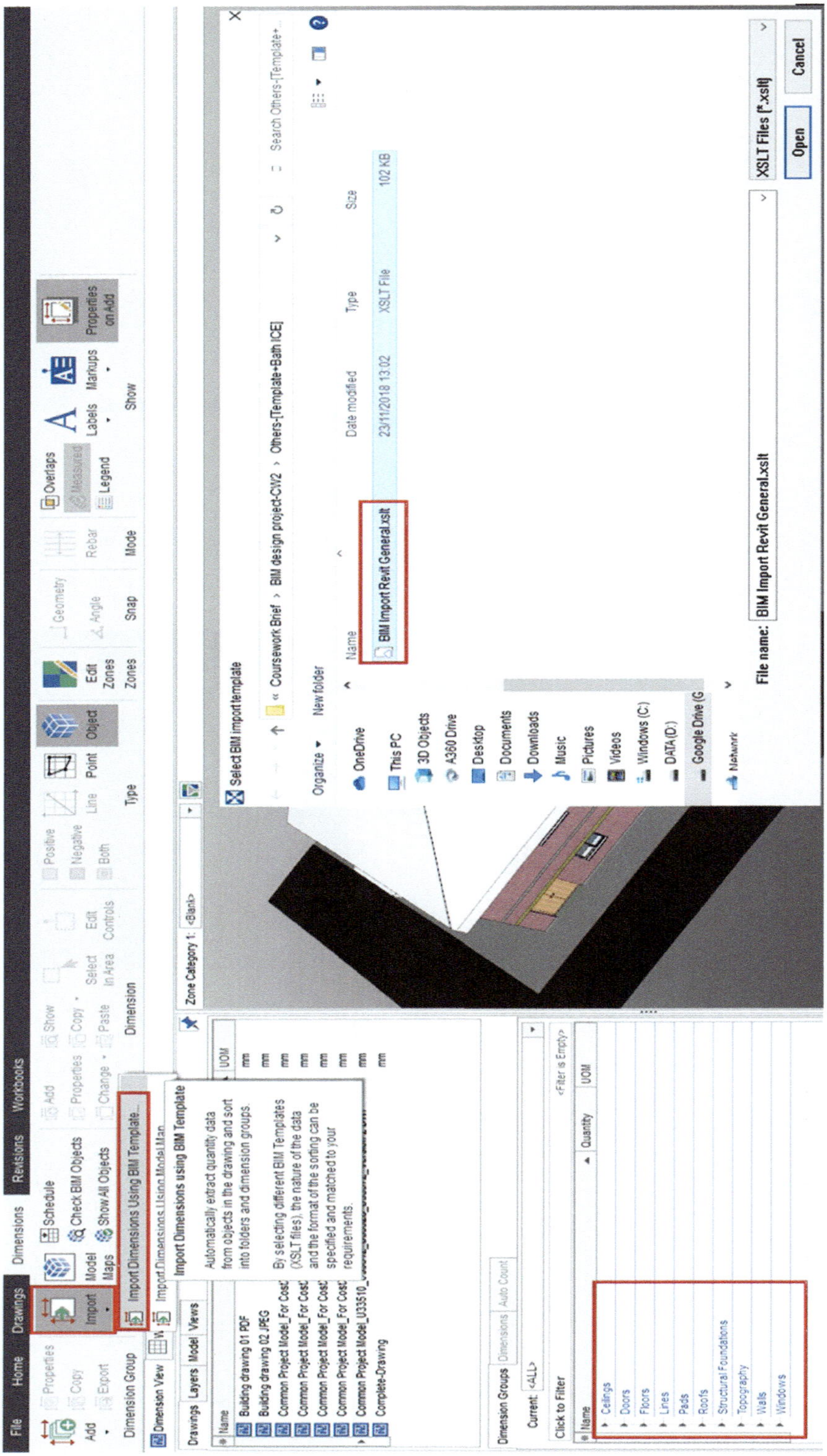

Figure 5.18 Model template QTO

Figure 5.19 5D BIM in Revit and iTwoCostX

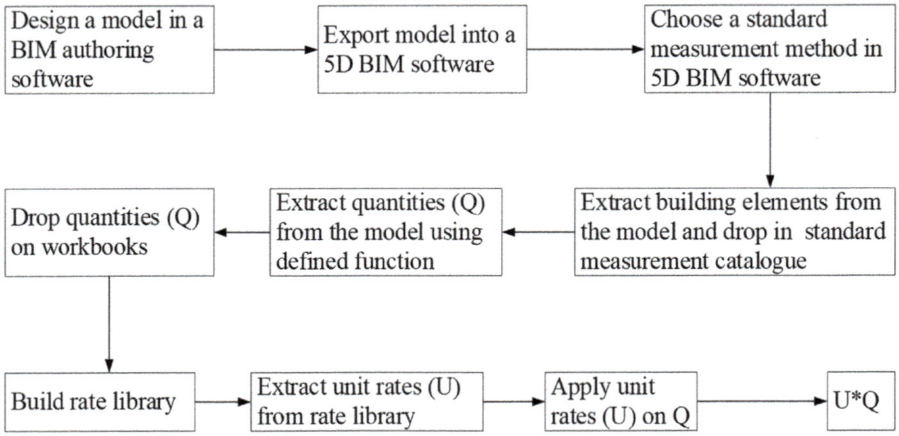

Figure 5.20 5D BIM workflow

Clearance: Two objects come within a specified distance. For example, this type of clash occurs when pipes need to have space for insulation around them. This type of clash occurs when an element is not given the spatial or geometric tolerances it requires or its buffer zone is breached.

Duplicate: Two objects are identical in types and in the same position. You can use duplicate to identify items in the scene that have been duplicated by mistake.

Each clash has a current status associated with it and each status has coloured icon to identify it.

The process of clash detection is presented in Figures 5.21 and 5.22.

Approved: A clash previously found and approved by someone.

Reviewed: A clash previously found and marked by someone as reviewed.

New: A clash found for the first time in the current run of the test.

Active: A clash found in a previous run of the test and not resolved.

Resolved: A clash found in a previous run of the test and not found in the current run of the test.

The generalized clash detection process is presented in Figure 5.23.

5.2.6 *BIM for Project Programme or Scheduling*

Traditionally, project programmes have been designed using software like MS Project, Asta PowerProject, Primavera, etc. These software systems offer the possibility of defining the different tasks including duration, start, and end dates. Developing project programmes in BIM software goes a bit further, by federating the geometric (3D) model with the non-geometric (time) data. This is known as 4D BIM.

The most common software for developing 4D BIM are Navisworks and Synchro. In this book, the demonstration will of 4D BIM will be done using Navisworks. Three methods will be used.

Method 1: Manual technique

In this method, tasks are manually edited in the TimeLiner and then attached to the different components of the geometric model. It is important to note that when attaching tasks to the

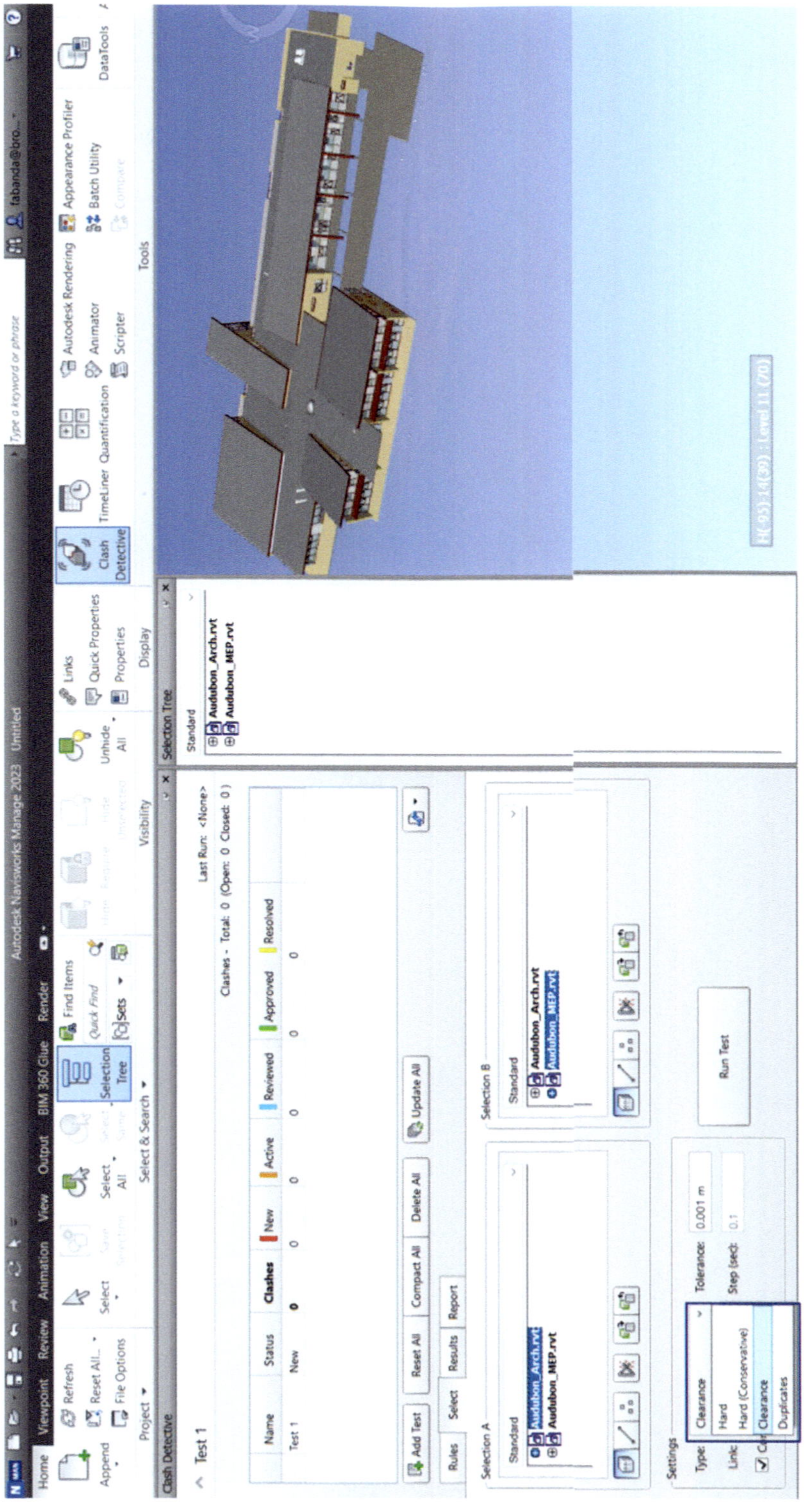

Figure 5.21 Clash detection in Navisworks

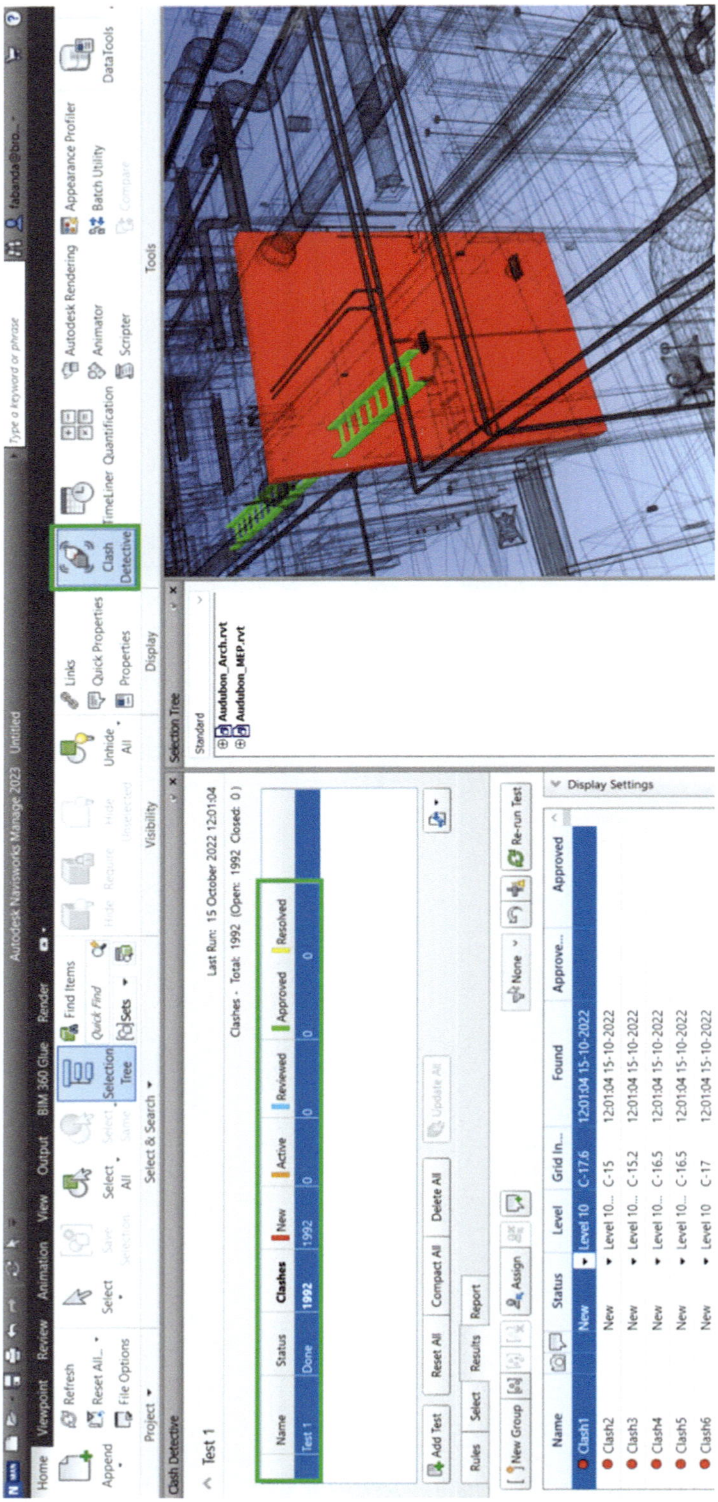

Figure 5.22 Clash detection status in Navisworks

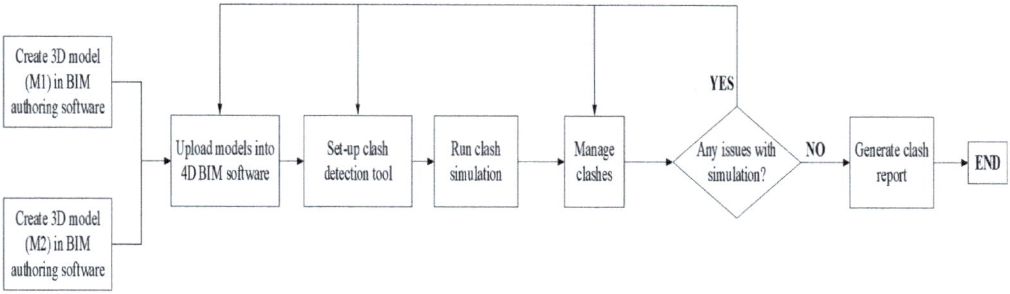

Figure 5.23 Clash detection workflow

different components of the building or project, it is advisable to explore directly the geometric model in the Scene view or the equivalent in the Selection Tree. The former is used when the component to be attached to the task is visible and easier to select from the Scene. If otherwise, then it is advisable to select the components from the Selection Tree and linked or attached to the relevant tasks in the TimeLiner.

Method 2: Semi-automatic

The first task in this method is to import the geometric model from a BIM authoring software, e.g., Revit. Secondly, the tasks are auto-generated in the TimeLiner using the Auto-Add Task tool. Once the Auto-Add Tasks is clicked, then the tasks are added taking into account only the top concepts of the hierarchy in the Selection Tree. The sub-concepts can then be manually edited, durations added and then attached to the geometric model.

The manual development of project programmes using Navisworks is presented in Figures 5.24 and 5.25.

Method 3: Automatic

In the automatic, the tasks are predefined in a traditional programming software, e.g., Asta Power Project while the geometric model is developed in a BIM authoring software, then both are imported into Navisworks and then federated.

The automatic development of project programmes using Revit, Navisworks, and Asta PowerProject is presented in Figure 5.26.

The generalized workflow for 4D BIM is presented in Figure 5.27.

5.2.7 BIM for Sustainability

In the built environment discipline, sustainability is defined as a 'process whereby firms utilize construction as a means of meeting their needs in a way that achieves value for money on a whole life basis in terms of generating benefits not only to the firms, but also to society and the economy, while minimizing damage to the environment.'

The initial EE/EC of a building is the energy expended in the extraction of its raw materials (indirect energy), as well as the energy used in the transportation of finished product to the job site and installation thereafter (direct energy) (Shrivastava and Chini, 2012).

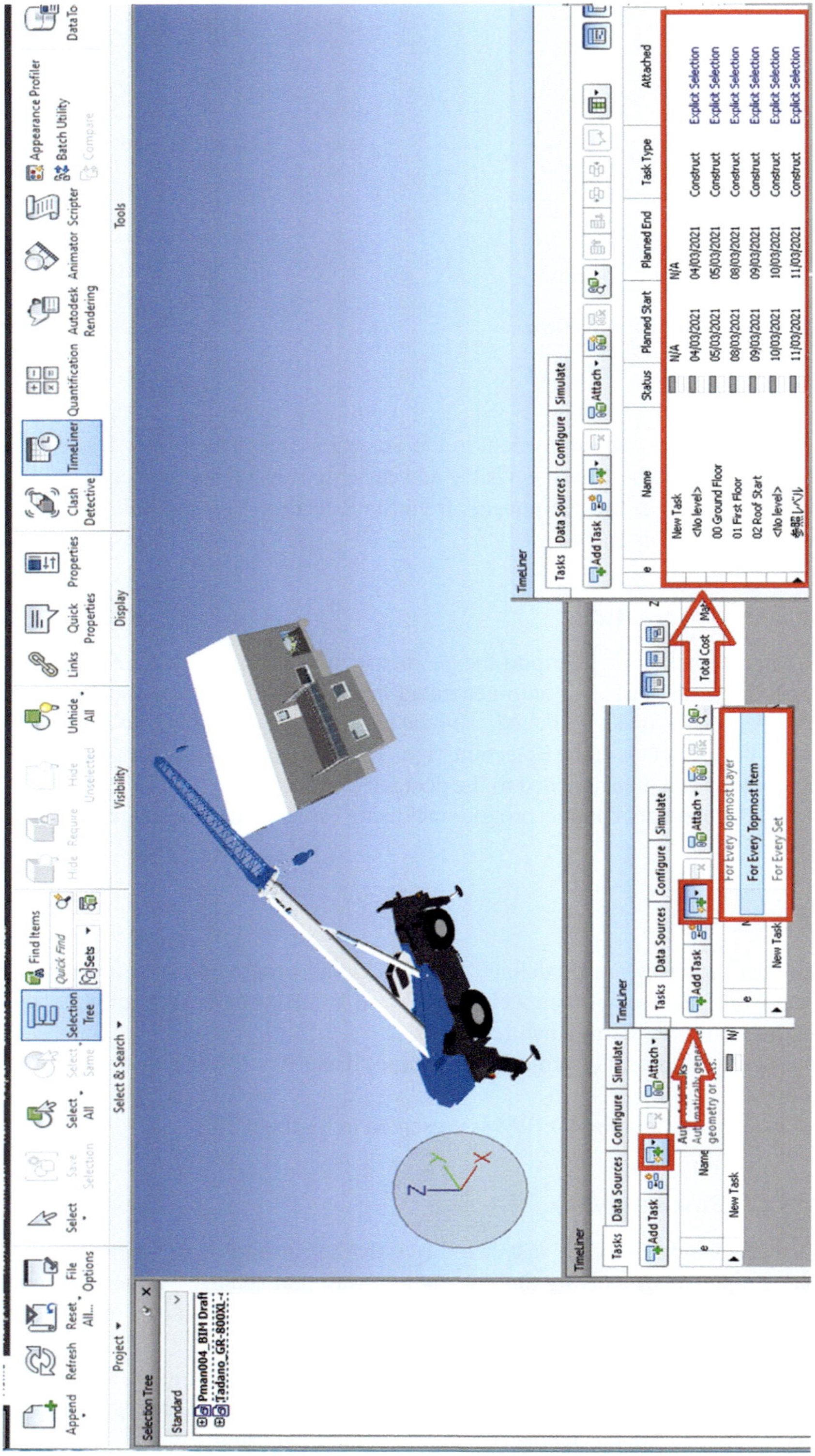

Figure 5.24 Developing project programmes in Navisworks

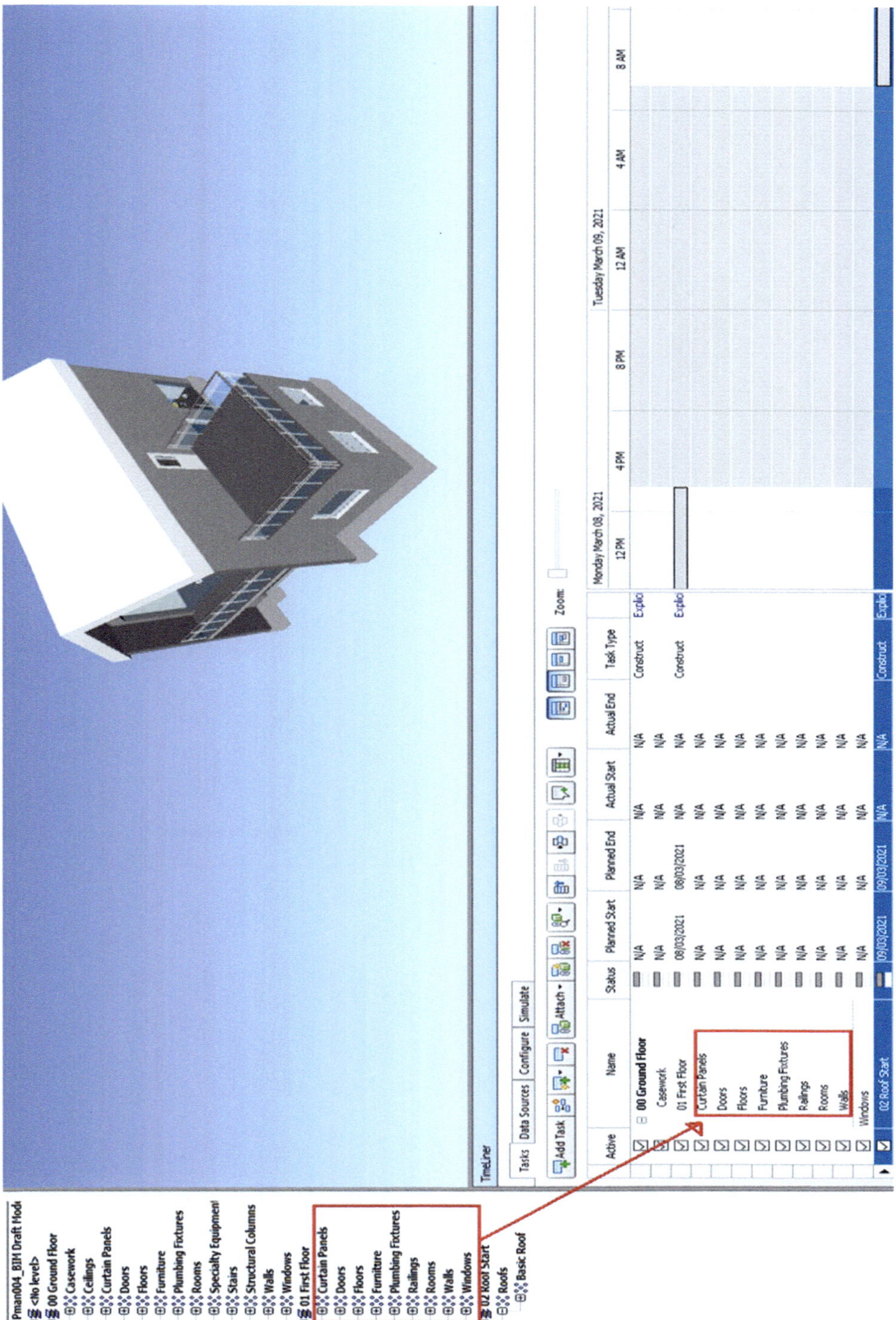

Figure 5.25 Sub-concepts are manually edited and attached to the geometric model

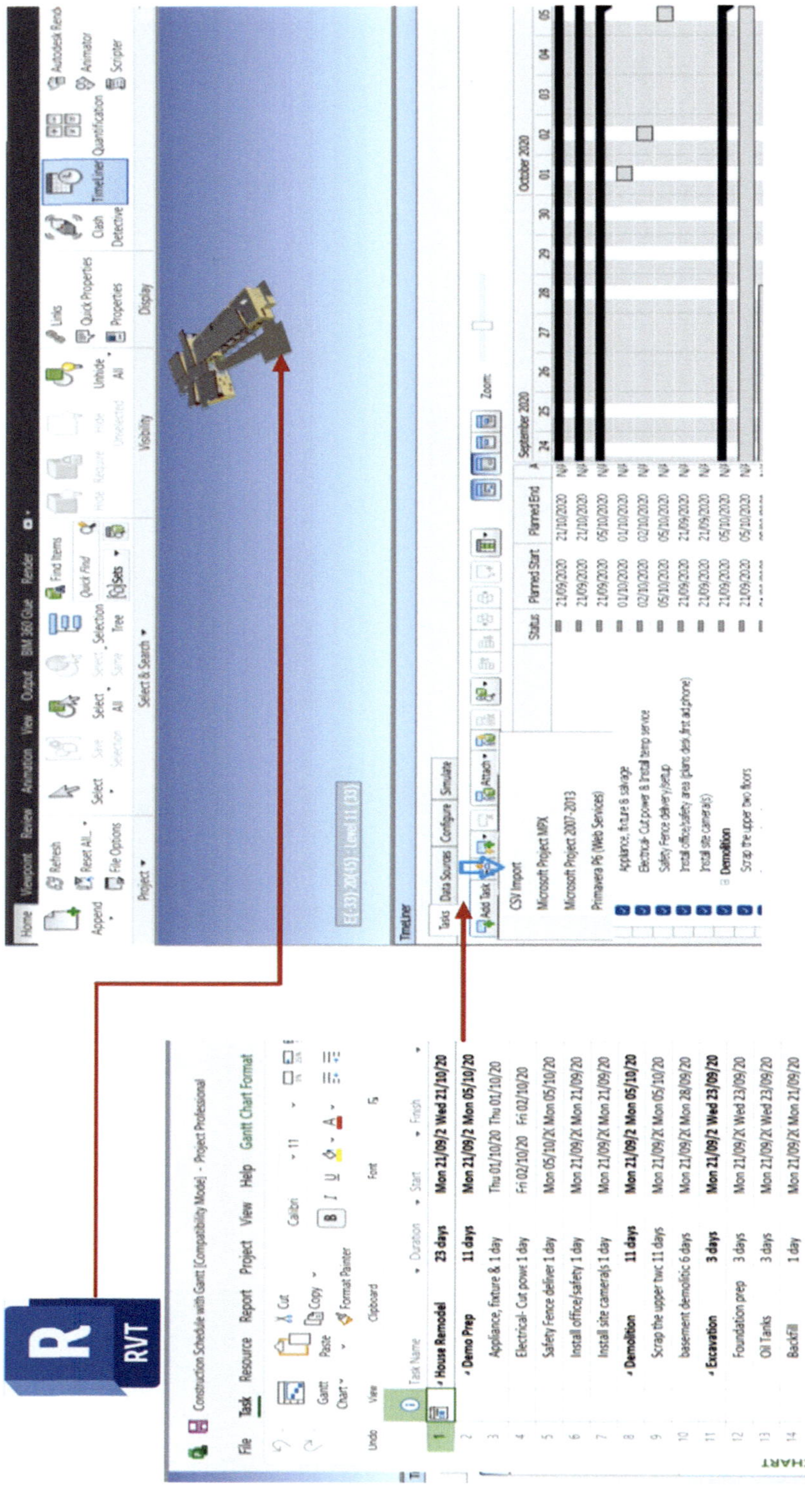

Figure 5.26 Semi-automatic generation of construction programmes

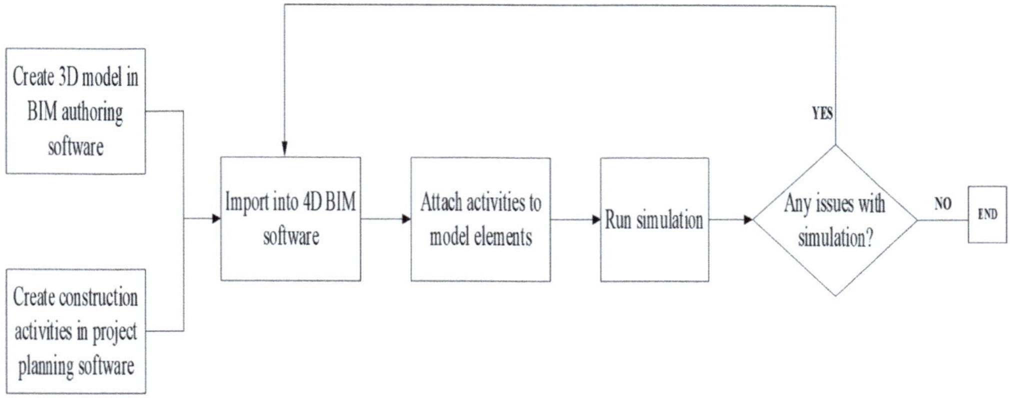

Figure 5.27 4D BIM workflow of project programmes

On the other hand, the recurring EE/EC of a building can be defined as energy expended to maintain, replace as well as recycle or dispose of it at the end of its useful life (Shrivastava and Chini, 2012).

OE is consumed in operating the buildings (Shrivastava and Chini, 2012).

Why have EE and EC suddenly become important?

To answer this question, it is important to revisit the well-established 'law of conservation of energy.' It states that the total energy of an isolated system remains constant; it is said to be conserved over time. In other words, energy can neither be created nor destroyed; rather, it can only be transformed or transferred from one form to another.

The total energy of a building is equal to the sum of its embodied and operational energy.

Let TBE, OE, and EE represent total building energy, operational energy, and embodied energy, respectively. Then:

$$\text{TBE} = \text{OE} + \text{EE} \tag{5.1}$$

The science of computing OE is well-known and can be easily applied to determine it. Also, with new building energy efficiency technologies, building regulations, home energy conservation acts, and other standards, it is possible to significantly reduce the operational energy of a building. Similar to the law of conservation of energy, it means at some point in time; the OE will be very small compared to EE. A study clearly illustrates this phenomenon (Figures 5.28 and 5.29).

The generalized BIM workflow embodied energy and carbon computation is presented in Figure 5.30.

5.2.8 *BIM for Retrofitting, Renovation, Refurbishment*

The terms retrofitting, renovation, and refurbishment will be explained in order to provide a context of how BIM improving the status of a building.

According to Design Buildings Wiki (DBW, 2022), the term '**renovation**' refers to the process of returning something to a good state of repair. In the built environment, **renovation** refers to the process of improving or modernizing an old, damaged, or defective building. This

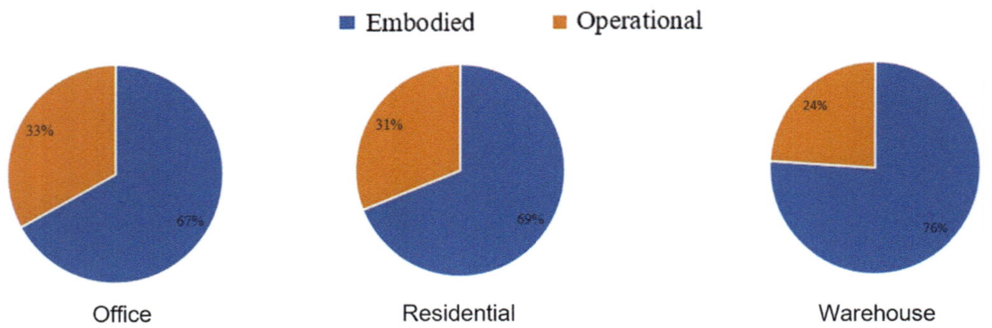

Figure 5.28 Indicative embodied vs operational carbon emissions for different building types (Source: RICS, 2017)

is as opposed to 'retrofitting' which is providing something with a component or feature not originally fitted, or 'refurbishment' which is a process of improvement by cleaning, decorating, or re-equipping. Refurbishment is the process of improvement by cleaning, decorating, and re-equipping. It may also include elements of retrofitting with the aim of making a building more energy-efficient and sustainable.

For BIM to be applied to the aforementioned activities, there must be an existing asset for which the owner wants to undertake changes or upgrade for various reasons including performance improvement. In practice, most existing buildings lack complete drawings that reflect the physical asset in reality. As such the asset, in this case a building can be laser scanned and the point cloud converted into a 3D BIM model. The process is presented in Figure 5.31.

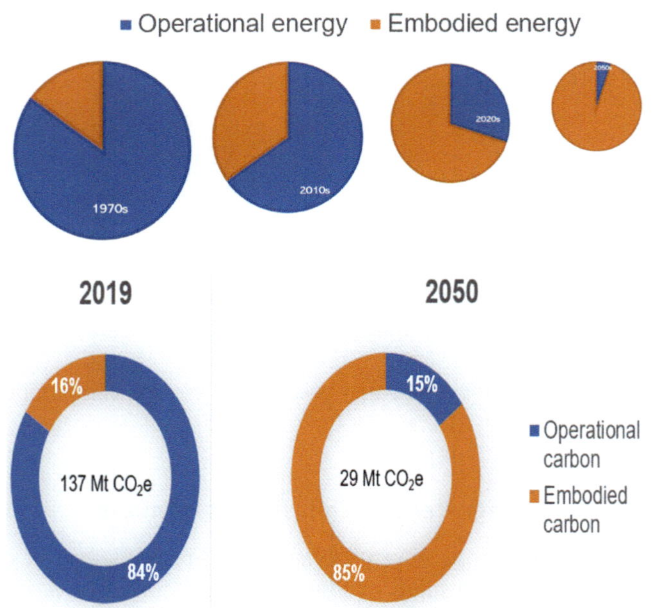

Figure 5.29 (a) Evolution of embodied vs operational energy over time (Source: (LETI, 2020)). (b) Carbon emissions from Australia's buildings in 2019 and 2050 (Process LCA) [GBCA and thinkstep-anz. (2021).]

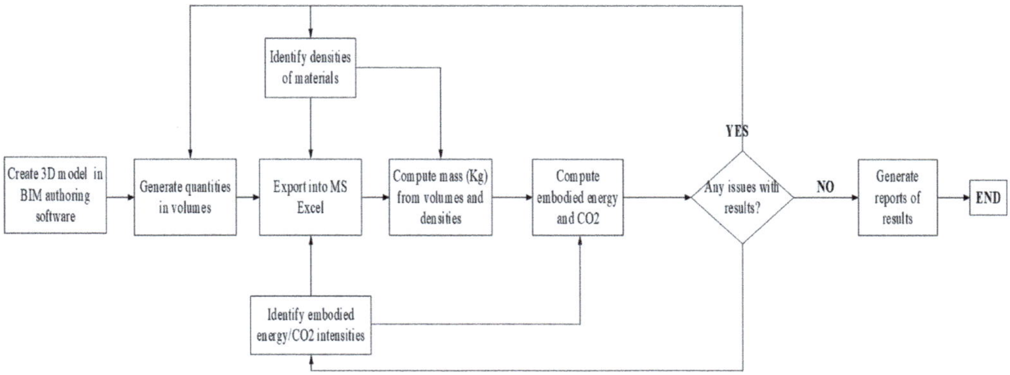

Figure 5.30 Embodied energy and CO_2 evaluation workflow

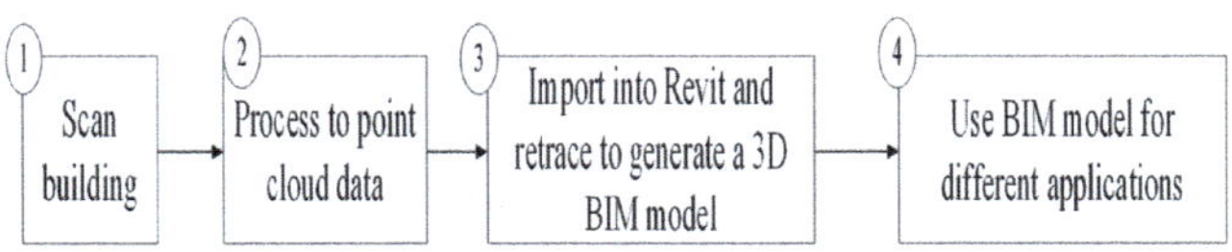

Figure 5.31 Scan-to-BIM process

This approach for developing or generating from scanning and cross-checking field data is called scan-to-BIM or field-to-BIM (Cropp, 2020). On the other hand, BIM-to-field is the mirror image of this process, where BIM data are downloaded validated on-site using a scanner and then fed back into the model (Cropp, 2020).

The process of scan-to-BIM involves processes that depend on the different software and hardware systems involved. A typical example is the scan-to-BIM workflow presented in Figure 5.32.

The scan-to-BIM or field-to-BIM process involved activities that are supported by software and hardware. Given the growing markets of digital construction technologies, the field-to-BIM process can be undertaken using different technologies. To facilitate understanding, a summary of the different field-to-BIM technologies will be presented in Table 5.1.

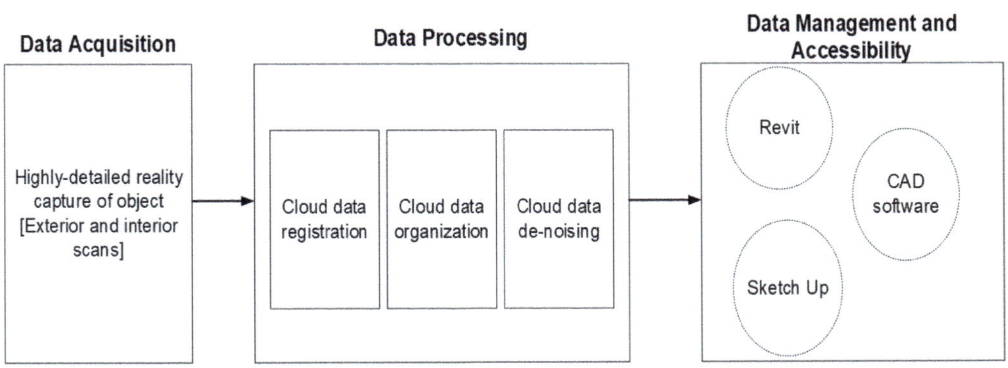

Figure 5.32 Specific scan-to-BIM workflow

Table 5.1 Relationship between hardware and software (input/output files) in the scan-to-BIM process

	3D-laser scanner	*Software*	*File in (input to software)*	*File out (output from software)*
3D laser scanner-dependent software	Trimble TX [Data output format (RWP)]	Trimble Realworks	RWP, XYZ, E57, LAS, LAZ, ZFS, RSP, FLS, DP, PTX, PTS	E57, ASC, LAS 1.2, LAS 1.4, LAZ, POD, PTS, PTX, TZF, BSF, KMZ, DWG, DXF, DGN, FBX, OBJ
	FARO [Data output format (FLS)]	FARO Scene	FLS, XYZ, CVS, COR, CPE	PTC, PTX, PST, XYZ, DXF, IGES, VRML, E57
	Leica [Data output format (PTX and PTS)]	Leica Cyclone	XYZ, PTS, PTX, LAS, E57, ZFS, DP	XYZ, PTS, PTX, E57, DXF, PCI/CWF, DBX, gbXML
3D laser scanner non-dependent software		LiDAR360	LiData, las, laz, asc, neu, xyz, pts, csv, ply	LiData, las, laz, asc, neu, xyz, pts, csv, ply
		PolyWorks	IGES, STEP, DXF, JT, OBJ, PLY, POL, STL, VRML 2.0	IGES, STL, DXF
		PointCab	dp, e57, asc, fws, lsproj, fls, las, laz, lsdx, lse, dae, 3ds, ifc, stl, mpc, ply, ptg, ptx, rps, pvtp, vtp, xyz, zfs, zfprj	dwg, dxf, dae, 3ds
BIM-based tools		Autodesk Recap	ASC, CL3, CLR, E57, FLS, FWS, ISPROJ, LAS, PCG, PTG, PTS, PTX, RDS, TXT, XYB, XYZ, ZFS, ZFPRJ, DXF, DWG	RCS, RCP, PCG, PTS, E57, DXF, DWG
		Autodesk Revit	DWG, RVT, DXF, gbXML, RCP, DWF	DWG, DXF, gbXML, FBX, DGN, ACIS, IFC
		Bentley	POD, OBJ, SHP, DXF, DWG, ESRI, E57, ZFS, LAZ, LAS, FLS, FWS, XYZ, PTS, PTX, PTZ, TXT, LWO, CL3, BIN, RSP, 3DD	POD, PTS, XYZ
		ArchiCAD	DWG, IFC, 3DS, 3DM, SKP, KMZ, STL, PDF, DXF	PDF, DWF, DXF, DWG, DGN
Web-based visualization and processing		FARO SCENE WebShare	e57, XYZ, PTX, and PTS	rcp, pod, xyz, e57
		Leica Pegasus	SHP, LAS, LAZ	SHP, LAS, LAZ
		Flyvast	LAS, LAZ, XYZ, PTX, PLY, OBJ	FLY, dxf, shp
		Voxxlr	IFC, Dxf, e57, pts, ply, las, laz	

5.2.9 BIM for Structures

In order to examine how BIM can be applied in structures, it is imperative to define structural engineering and structure. *Structural engineering* is the study of how various components of a building or other structure act together to transmit forces down to the foundations. A *structure* is an organized combination of connected parts designed to carry loads and provide adequate rigidity.

The stages of structural engineering are as follows:

- *Structural planning stage:* This is when a structural scheme is devised to suit both the purpose of the building and the site conditions.
- *Structural analysis stage:* This is when the loads on the structure are determined and the way the loads disperse through the structure is analysed using the principles of *structural mechanics*.
- *Structural element design stage:* This is when the size and properties of each member are determined.
- *Structural detailing stage:* This is when detailing drawings are produced to illustrate how the structure is to be constructed on-site.
- *Structural specification stage:* This is when the specification clauses are compiled to define the standard of materials and workmanship to be used.
- *Construction stage:* This is when the structure is built, with appropriate supervision inspection and testing to ensure that it complies with the drawings and the specification (Figure 5.33).

5.2.10 BIM for Collaboration

BIM for collaboration transforms the traditional AEC industry by breaking down silos, promoting effective communication, and optimizing the construction process. It empowers stakeholders to work in a coordinated and efficient manner, resulting in improved project outcomes, reduced errors, and enhanced cost-effectiveness throughout the entire project lifecycle.

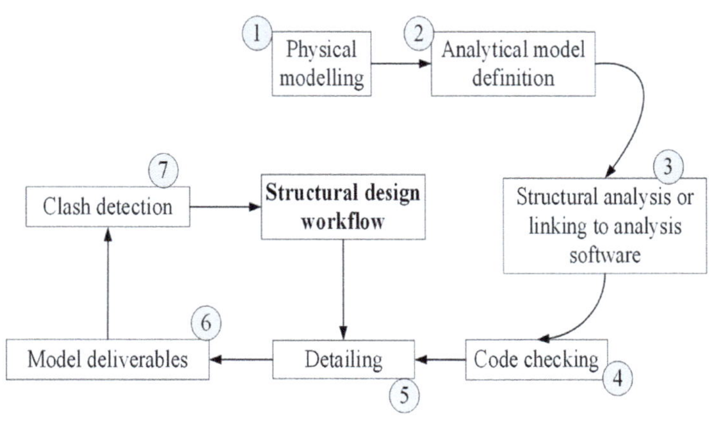

Figure 5.33 Simplified BIM workflow for structural engineering design (Adapted from: Tushaj, 2021)

5.3 Optimizing BIM Productivity Through Integrating Multiple Software for Practical Applications

The world of digital technologies is experiencing a surge in innovative technologies with many being from common and leading vendors such as Microsoft. Despite the popularity of Microsoft systems and their uses in many sectors, their applications in the built environment are hard to find. The ensuing sections will focus on the integration of BIM and Microsoft systems for different construction applications.

The two main Microsoft products to be considered are Microsoft Power BI and Microsoft Power App. Power BI is an interactive data visualization software that primarily focuses on business intelligence. Microsoft Power App is an application for the easy development of mobile and web apps for any business need by those with or without any technical or development experience.

Also, there are some innovative technologies in some emerging areas that can also significantly improve information management in construction practice, for example, Autodesk Tandem and Azure Digital Twins for Digital Twin, TX Laser Scanner for laser scanning, and Polycam for mobile laser scanning.

Autodesk Tandem is a cloud-based digital twin technology platform. It enables projects to start digital and stay digital, transforming rich data into business intelligence. By harnessing BIM data throughout the process, AEC firms can create and handover a digital twin to building owners and operators. The easily accessible, contextual, and insightful data they receive makes for ready-to-go operations.

5.3.1 Revit-Power BI Applications

In Revit quantities and schedules of items can be generated in CSV format which can be converted into MS Excel. Each of these file formats can be imported into Power BI and used in developing various applications. A typical workflow is presented in Figure 5.34.

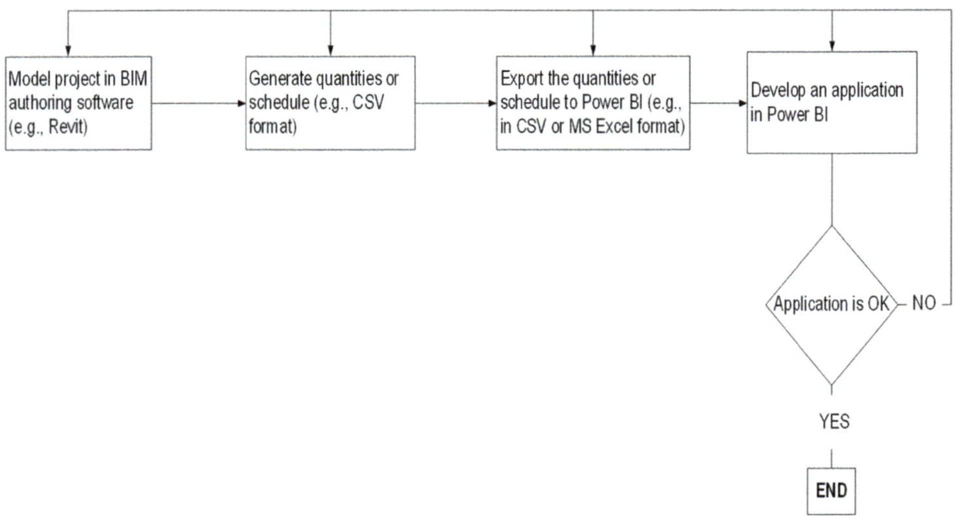

Figure 5.34 Revit-Power BI Workflow

To facilitate understanding the implementation of the workflow in a software environment of Figure 5.34 will be presented in Figure 5.35. To facilitate readers' understanding of the potential of Power BI, a typical application in construction, i.e., generating of schedules, will be illustrated.

With the schedule in Power BI, different visuals can be developed to demonstrate the different building applications. For example, it is possible to illustrate the number of windows in the different rooms and levels of the buildings.

5.3.2 Revit-Navisworks Manage-Power BI Applications

There many applications that can be developed through the combination of Revit, Navisworks, and Power BI. The first application is about quantity take-offs. The workflow for quantity take-offs using Revit-Navisworks-Power BI is presented in Figure 5.36.

The second application is the combination of Revit-Navisworks Management-Power BI for the development of project programmes. The third application that can be performed through the combination of combination of Revit-Navisworks Management-Power BI is clash detection (Figure 5.37).

Similar to the first example, the workflows of the project programmes and clash detection of the second and third applications can be developed and implemented in this case is presented in Revit-Navisworks Management-Power BI.

5.3.3 Revit-Autodesk Construction Cloud (ACC)-Power BI Applications

Many applications can be developed in Power BI from models developed in Revit. Similar to Section 5.3.1, a Revit model can be exported to ACC and used for many applications such as document management, issue tracking, asset management, health and safety, etc. From ACC, the data can be converted into an Excel file and then exported to Power BI where other intelligent applications can be developed for easy interrogation and visualizations by non-expects.

The workflows for issue tracking and asset management applications are presented in the ensuing Figures 5.38 and 5.39.

5.3.4 Revit-ACC-Power App Applications

Similar to Revit-ACC-Power BI, in the case of Power App, it comes with a model developed in Revit, then exported to ACC, and then converted into MS Excel. From MS Excel, it is stored in Power App accessible formats such as OneDrive, Google Drive, SharePoint, SQL Server, and Microsoft Dataverse. Once stored in any of the aforementioned platforms, Power App can then be used to read the data and an application developed depending on the intended purposes (Figure 5.40).

5.3.5 Revit-Autodesk Tandem-Power BI Applications

The connection among the three systems commences with a model developed in Revit, and then, Autodesk Tandem is used in importing it. From Autodesk Tandem, the model is converted to MS Excel file which is then exported into Power BI. In Power BI, the file will then serve as the basis to develop practical applications.

The workflows for the principle and implementation are presented in Figure 5.41.

An example of how to generate an inventory of all the elements in the project is presented in Figure 5.42.

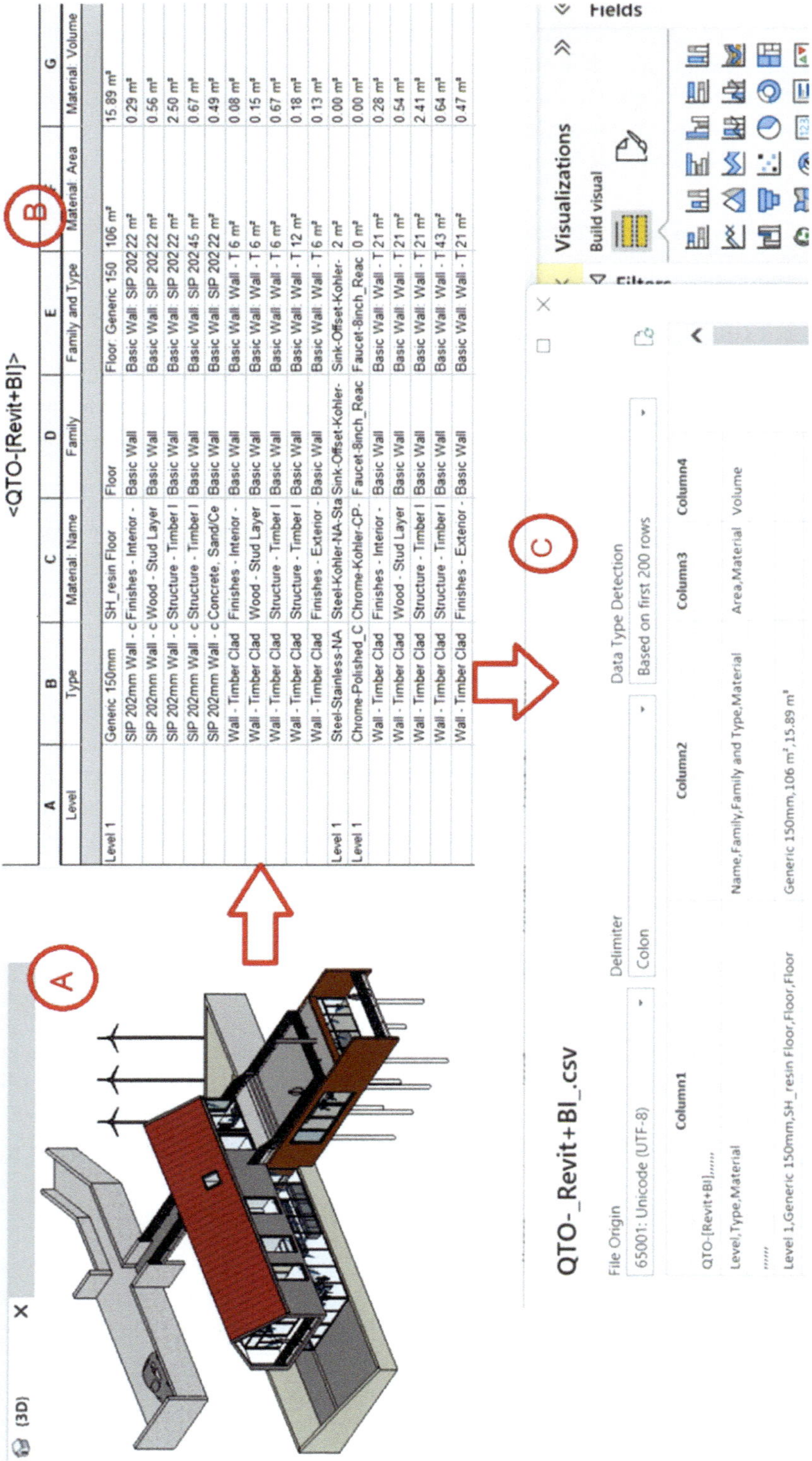

Figure 5.35 Implementation of scheduling in Revit-Power BI software environment

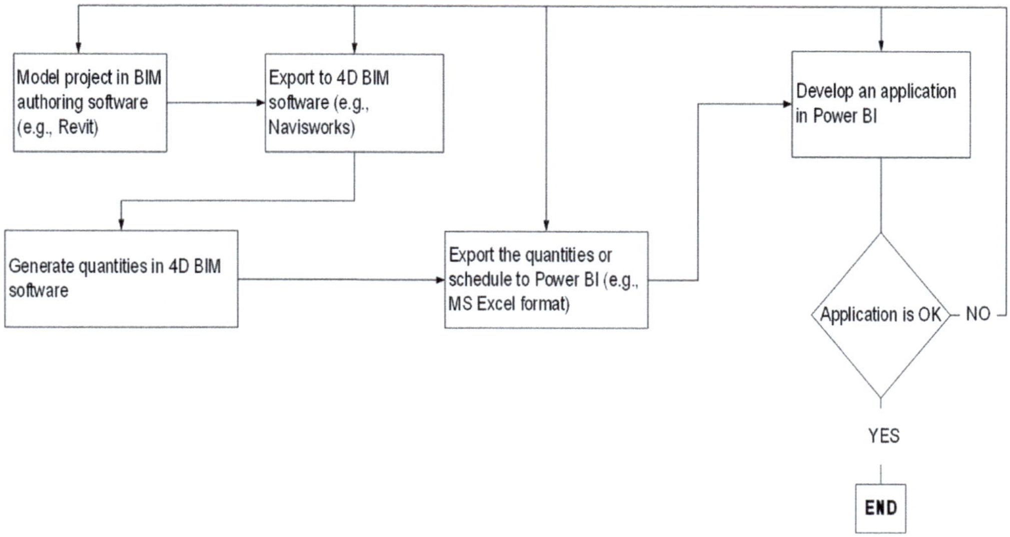

Figure 5.36 Quantity take-off workflow in Revit-Navisworks Manage-Power BI

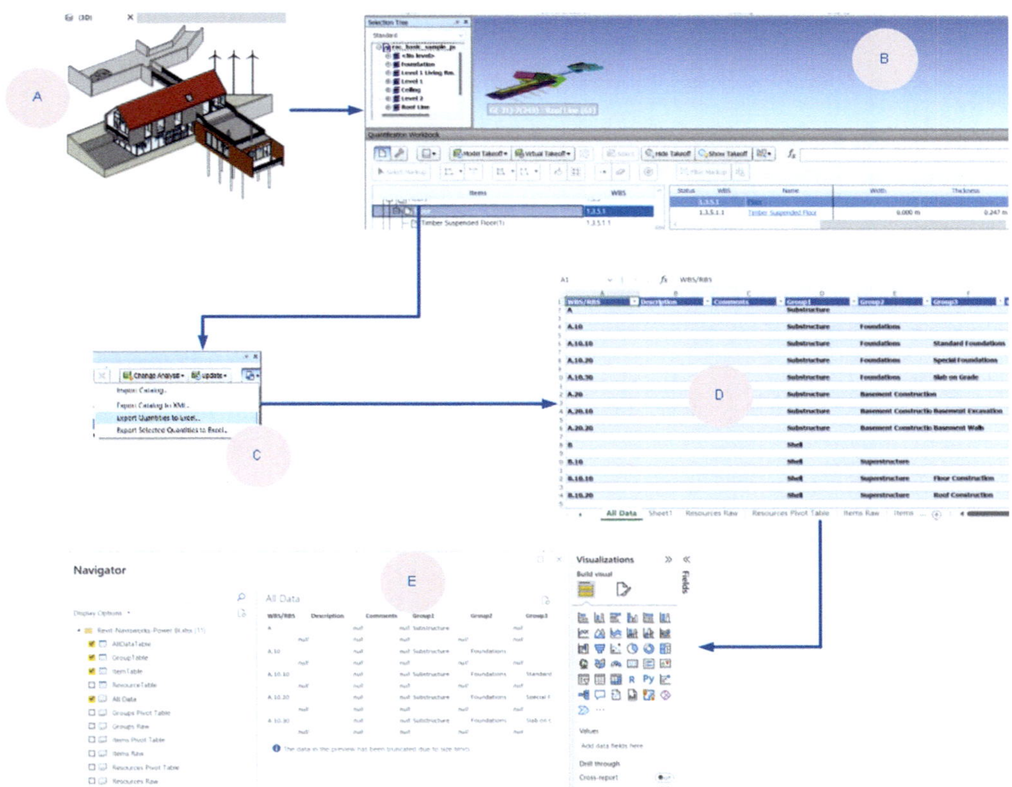

Figure 5.37 Implementation of QTO workflow in Revit-Navisworks Manage-Power BI environment

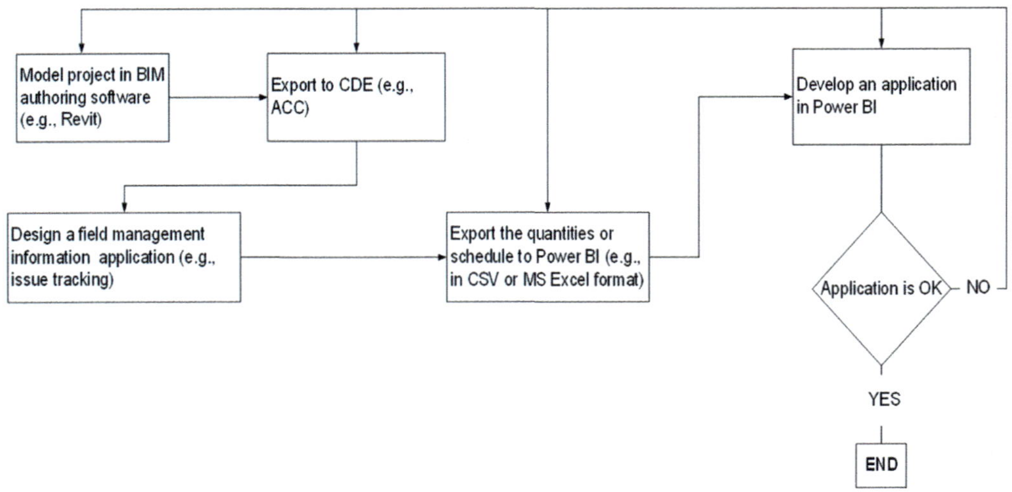

Figure 5.38 Issue Tracking in Revit-ACC-Power BI

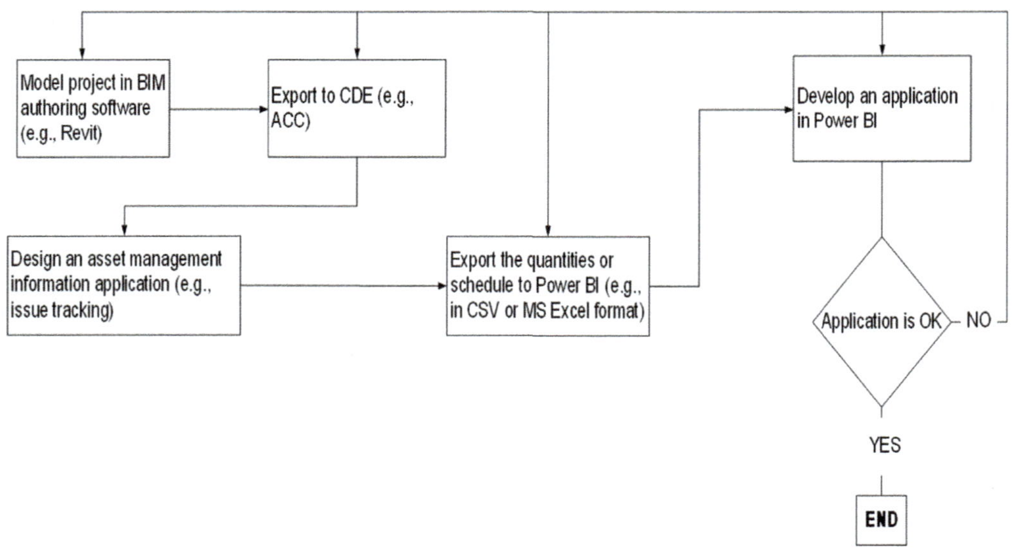

Figure 5.39 Asset Management in ACC-Power BI

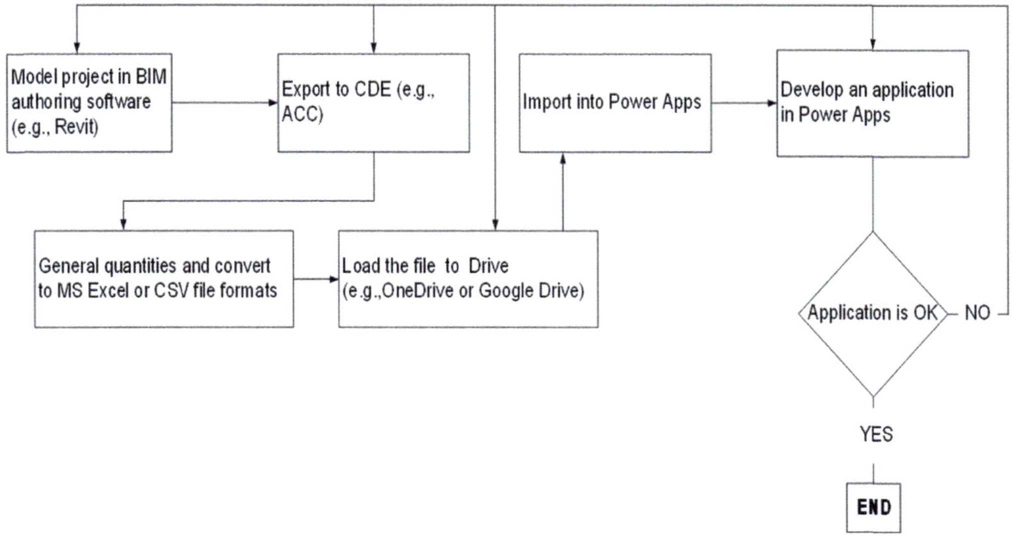

Figure 5.40 Revit-ACC-Power Apps Workflow

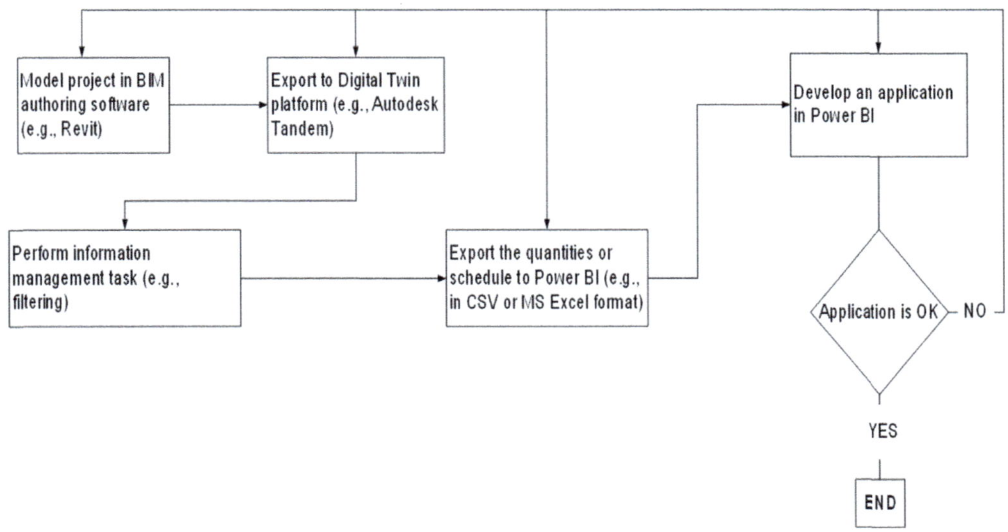

Figure 5.41 Revit-Autodesk Tandem-Power BI

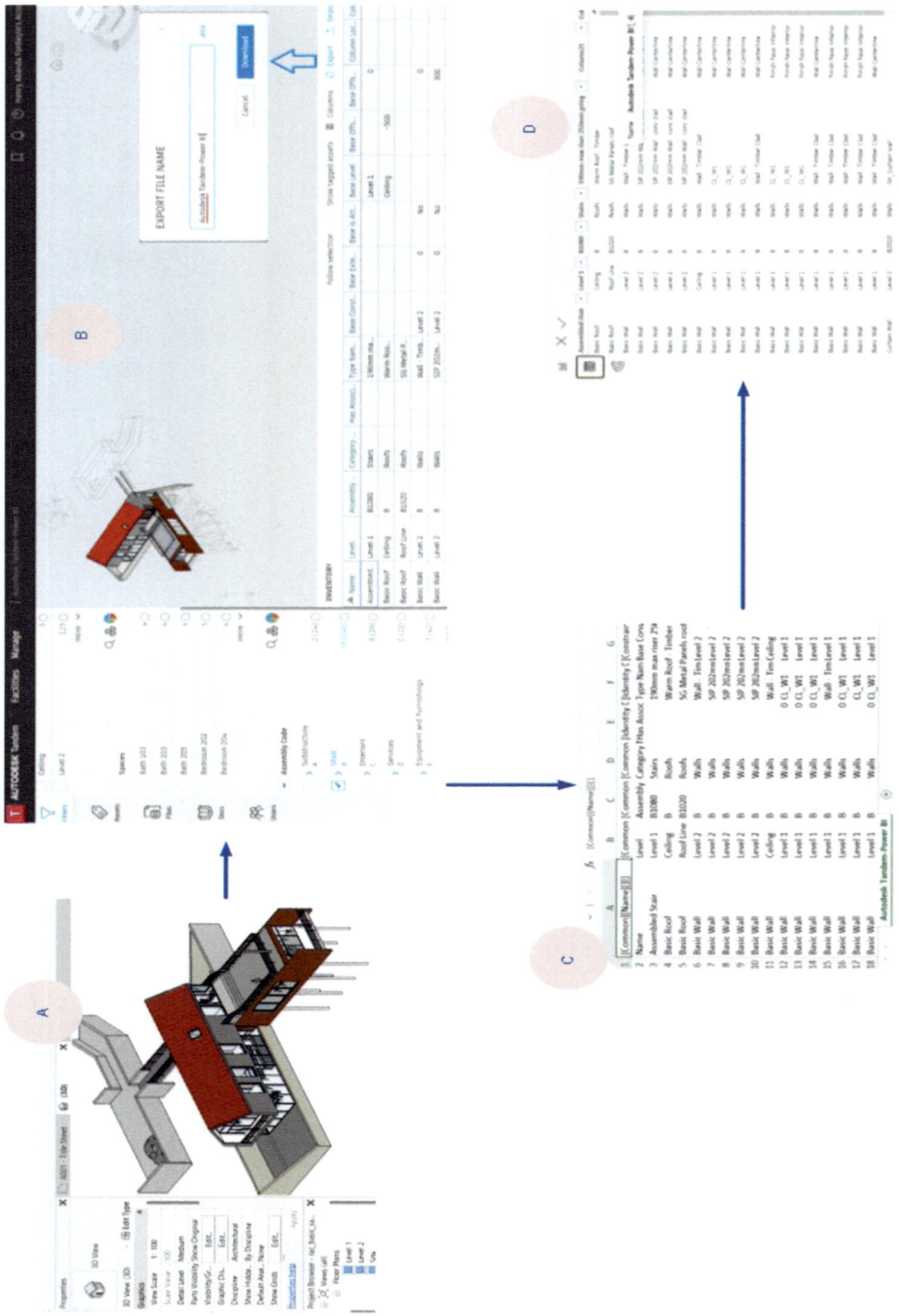

Figure 5.42 An inventory of components or elements in the project

Figure 5.43 Scan-to-BIM application

5.3.6 Scan-to-BIM Applications

This is about scanning assets and converting them into BIM format. This can be done by using 3D laser scanners or mobile app scanners and the output (point cloud) is then used to generate BIM models for different applications. The workflows differ and depend on the software systems involved (Figure 5.43).

5.3.7 Visual Programming Applications

Visual programming in construction offers the potential to streamline processes, automate repetitive tasks, and enhance collaboration and decision-making. By simplifying complex programming concepts through visual elements, construction professionals can leverage the power of technology more effectively, leading to improved efficiency, accuracy, and innovation in the industry.

- BIM: Visual programming can enhance the implementation of BIM workflows in construction. BIM involves creating digital representations of buildings and infrastructure, and visual programming can be used to create custom scripts or workflows that automate tasks, generate reports, or extract specific information from BIM models.
- Parametric Design and Generative Architecture: Visual programming enables parametric design, which involves creating models or designs that respond to specific parameters or constraints. Architects and designers can use visual programming tools to generate complex geometric forms, optimize building performance, and explore design alternatives based on defined parameters.
- Design Optimization and Analysis: Visual programming can be employed to automate design optimization and analysis processes. It allows construction professionals to define design parameters, run simulations, perform analysis, and evaluate multiple design scenarios rapidly. This helps in identifying the most efficient and cost-effective design solutions.
- Prefabrication and Modular Construction: Visual programming tools can streamline the design and fabrication of prefabricated or modular components in construction. These tools enable the creation of custom workflows that automate the generation of shop drawings, assembly instructions, and production data for off-site manufacturing. Visual programming can enhance coordination between design teams, manufacturers, and construction crews involved in prefabrication processes.
- Construction Sequencing and Simulation: Visual programming can assist in the planning and simulation of construction sequences. By using visual programming tools, construction teams can create visual representations of construction processes, simulate construction sequences, and evaluate potential conflicts or clashes. This helps in identifying potential issues and optimizing construction schedules.

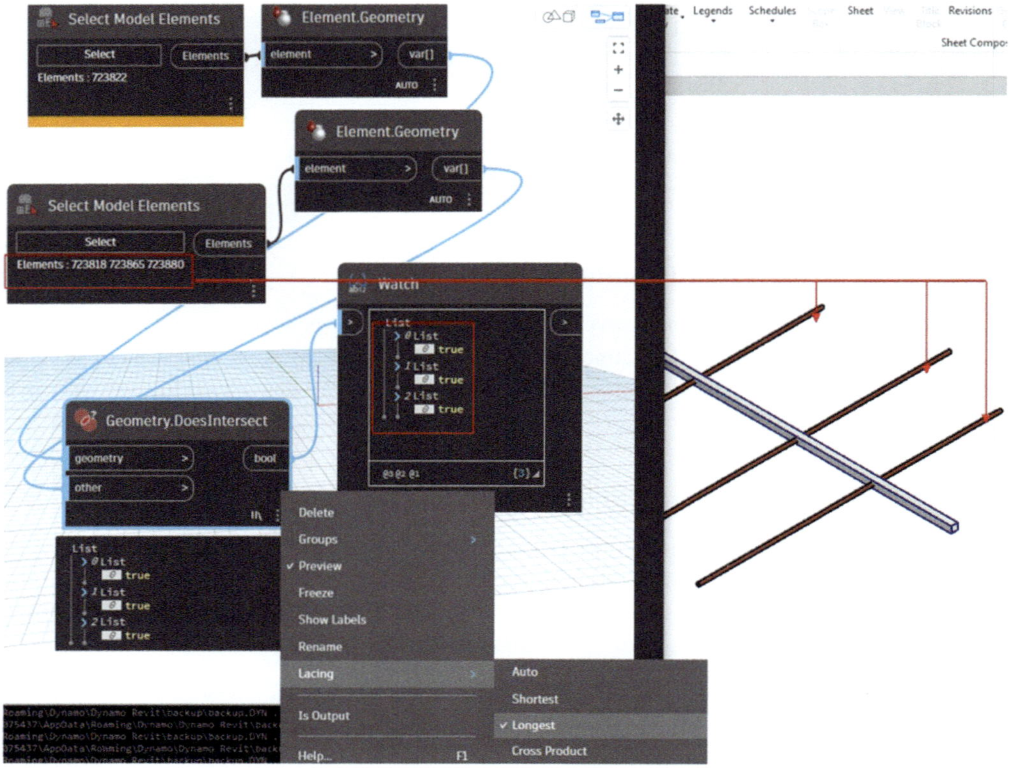

Figure 5.44 Dynamo-Revit Application-Clash detection

- Project Management and Data Visualization: Visual programming tools can be used to develop customized project management dashboards and data visualization applications. These applications can integrate data from various sources, such as project schedules, cost data, and quality control information, and present it in a visual and interactive manner, providing real-time insights into project progress, resource allocation, and performance analysis (Figure 5.44).

5.4 Conclusions

The exploration of Digital Construction Applications in this chapter marks a significant stride towards understanding the practical integration of digital technologies within the construction industry. Through detailed process maps and software implementations, we have unveiled the intricacies of digital construction technologies, showcasing their pivotal role in enhancing project execution, collaboration, and efficiency. This examination not only highlights the technological advancements in construction but also underscores the importance of adopting digital solutions for a competitive edge in today's rapidly evolving construction landscape.

As we transition from the theoretical underpinnings and practical applications of digital construction technologies, our journey leads us to the next frontier: 'Cloud BIM and CDEs.' This upcoming chapter delves into the transformative potential of cloud-based BIM technologies and the strategic implementation of CDEs for information management. With a focus on BIM 360's modules – Document Management, Field Management, and Asset Management – readers will gain insights into how cloud BIM facilitates unprecedented collaboration, data accessibility, and project management efficiency. These modules exemplify the dynamic capabilities of cloud-based platforms in streamlining project workflows, enhancing communication, and improving decision-making processes.

In the forthcoming chapter, we will explore the realms of Cloud BIM and CDEs, pivotal elements in the digitalization of the construction industry. This discussion will illuminate how cloud-based BIM solutions, particularly through BIM 360, revolutionize information management across the project lifecycle. By dissecting the functionalities and applications of Document Management, Field Management, and Asset Management modules, the chapter aims to provide a comprehensive overview of how these tools empower stakeholders to manage projects more effectively, ensuring that the right information is accessible to the right people at the right time. Through practical examples and case studies, we will demonstrate the operational efficiencies and collaborative synergies enabled by Cloud BIM and CDEs, offering readers a roadmap to harnessing these technologies for enhanced project outcomes.

Notes

1 www.nationalbimlibrary.com
2 www.BIMobject.com
3 www.bimstore.co.uk
4 https://www.revitcity.com/index.php

Chapter 6

Cloud-Based BIM

6.1 Introduction

The advent of BIM within the digital construction sphere has revolutionized the way industry professionals approach project management, design, and collaboration. As we progress further into the digital age, the significance of leveraging cloud-based technologies and CDEs has become increasingly paramount. Chapter 6 embarks on an exploratory journey into the realms of Cloud BIM and CDEs, key components that facilitate enhanced information management and collaborative efficiency across the construction project lifecycle.

In this chapter, we delve into the intricacies of Cloud BIM, examining how it extends beyond traditional BIM practices by offering a more dynamic, accessible, and collaborative approach to project management and execution. Through the lens of BIM 360 and its various modules – Document Management, Field Management, and Asset Management – we will uncover the practical applications and benefits of cloud-based solutions in managing complex construction projects.

Furthermore, the chapter will highlight the role of CDEs as the backbone of effective digital collaboration, enabling stakeholders to access, share, and manage project information seamlessly. Readers will gain a comprehensive understanding of how Cloud BIM and CDEs not only streamline workflows but also pave the way for a more integrated, transparent, and efficient construction industry.

Through a detailed exploration of these technologies, this chapter aims to equip readers with the knowledge and insights needed to navigate the evolving landscape of digital construction, fostering a deeper appreciation for the potential of Cloud BIM and CDEs in driving forward industry standards and practices.

6.2 Overview of Cloud BIM

Cloud BIM, or Cloud-based BIM, refers to the utilization of BIM technology over cloud computing platforms (Abanda et al., 2018; Nhachi and Abanda, 2017). This approach enables the storage, management, and sharing of BIM data on remote servers, accessible via the internet. Cloud BIM offers enhanced collaboration among project stakeholders, allowing real-time access to updated project information, models, and documentation from any location. It supports a more integrated and efficient project management process, facilitating seamless communication, data exchange, and decision-making across the construction project lifecycle.

DOI: 10.1201/9781003476740-6

6.3 Mobile BIM

Mobile BIM, or mobile building information modelling, refers to the use of BIM technology on mobile devices such as smartphones and tablets. BIM is a digital representation of the physical and functional characteristics of a building or construction project. It allows architects, engineers, construction professionals, and other stakeholders to collaboratively design, plan, construct, and manage buildings and infrastructure in a 3D virtual environment.

6.4 Common Data Environments for Information Management

A typical construction project consists of three main phases: design, construction, and operations. The design phase is vital. Traditionally, architects start with the conceptual design of the construction by working with the client requirement. HVAC engineers and MEP engineers participate later in the design process based on the design requirements of the architect. Once the conceptual design is converted to a construction-ready model, the actual construction kicks off. Project managers will get involved and the cost and schedule are managed accordingly. The problem with the above model is unforeseen surprises. A slight miscalculation from the structural engineers could go unnoticed until construction is started. This could cause rework, causing an unnecessary increase in cost. It will also impact the schedule of the project. On the other hand, project costing and schedule management are also rapidly adjusted during the construction phase to make the high demands of the client. Usually, 50–70% of the capital expenditure is committed during the conceptual phase of the project and this leaves very little room for project managers to adopt to ever changing project demands. The CDE provides a complete solution by giving all stakeholders a view of 'one truth' with integrated information.

Functions of a CDE are to ensure data integrity, provide a single source of truth, enhance collaboration among stakeholders, manage project documentation efficiently, and support decision-making processes by offering real-time access to up-to-date project information.

Container ID: A unique identifier assigned to a container within a CDE that houses project documents and data, facilitating organization, and retrieval.

Revision: Indicates the versioning of documents or files within the CDE, tracking changes and updates to ensure stakeholders access the most current information.

Status: Reflects the current state of a document or file within the CDE workflow, such as draft, under review, approved, or issued, guiding users on its readiness for use.

Classification: A systematic arrangement of data and documents based on predefined standards or categories, enhancing searchability and compliance with project or industry-specific requirements.

Centralization: A single source of information for the project, ensuring that everyone accesses the same data.

Accessibility: Data and documents can be accessed from any location, improving collaboration among stakeholders.

Security: Robust protocols to protect sensitive project information, including access controls and encryption.

Interoperability: Supports various file formats and software, facilitating seamless data exchange.

Version Control: Tracks revisions and updates to documents, ensuring the latest information is always available.

Compliance: Adheres to industry standards and regulations for data management and security.

Integration: Can integrate with other tools and platforms for project management, design, and analysis.

Scalability: Capable of supporting projects of varying sizes and complexities. Some selected examples of CDEs are presented in Table 6.1.

Table 6.1 Some selected common data environments

Name	*Provider*	*Website*
Asite		https://www.asite.com/
Autodesk Construction Cloud	Autodesk	https://construction.autodesk.co.uk/
BIMXtra	Clearbox	https://www.clearboxbim.com/products/bimxtra
BCDE	Bentley	https://www.bentley.com/software/bcde/
planBIM		https://planbim.io/
Plannerly		https://plannerly.com/
GliderBIM		https://glidertech.com/gliderbim/
Aconex	Oracle	https://www.oracle.com/construction-engineering/aconex/
BIMcollab		https://www.bimcollab.com/en/
usBIM		https://www.acca.it/bim-management-system
revitzo		https://revizto.com/en/
KROQI		https://kroqi.fr/en/
BIMPLUS	AllPlan	https://www.allplan.com/products/allplan-bimplus/
Procore	Procore	https://www.procore.com/en-gb
Trimble Connect	Trimble	https://connect.trimble.com/
Catenda Hub	Catenda	https://catenda.com/bim-solutions-open-standards/catenda-hub-common-data-environment/
Trimble Viewpoint	Trimble	https://www.viewpoint.com/en-gb/
3D Repo	Asite 3D Repo Ltd	https://3drepo.com/
Projectwise	Bentley	https://www.bentley.com/software/projectwise/
ProjectSight		https://projectsight.trimble.com/
ThinkProject		https://thinkproject.com/products/tpcde/
Zutec		https://www.zutec.com/
NewForma		https://www.newforma.com/
InEight		https://ineight.com/
Dalux		https://www.dalux.com/solutions/information-management/
BIMsync		https://bimsyncmanager.firebaseapp.com/home
BIM Track		https://www.vrex.no/wp-content/uploads/2021/09/Bimtrack_heart_Vrex.png
Archdesk		https://archdesk.com/
BEXEL Manager		https://bexelmanager.com/
Bricsys 24/7		https://www.bricsys.com/en-gb/247

6.5 BIM 360 Tools

6.5.1 *BIM 360 Document Management*

This is a module within Autodesk's BIM 360 platform that provides a cloud-based solution for managing construction documents and drawings. It facilitates the organization, distribution, and sharing of project documents among team members, ensuring that everyone has access to the most current information. This module supports version control, document review, and approval workflows and offers tools for viewing, marking up, and annotating documents directly within the web interface, enhancing collaboration and reducing the risk of errors and miscommunications throughout the project lifecycle.

Create a company≫Add actors≫then add actors to a project. Then, you can add create your folders ≫ then add projects or model.

6.5.2 *BIM 360 Field Management*

This module focuses on field data management, offering tools for quality, safety, and commissioning checklists. It enables users to create, manage, and resolve issues on-site, facilitating real-time communication between field and office.

6.5.3 *BIM 360 Design Collaboration*

BIM 360 Design Collaboration: Designed for collaboration in BIM projects, this module allows team members to co-author Revit models in real time, manage deliverables, and track project progress, enhancing coordination and reducing conflicts.

6.5.4 *BIM 360 Model Coordination*

BIM 360 Model Coordination: Automates clash detection and helps teams resolve conflicts before they occur on the construction site. It supports model aggregation and coordination, streamlining workflows for identifying and addressing design and constructability issues early in the project lifecycle. The focus of model coordination is a clash coordination. To facilitate understanding, the traditional workflow is presented in Figure 6.1.

6.5.4.1 *Traditional Workflow*

The following weaknesses can be noted about the traditional workflow of model coordination.

Reliance on one individual

- A potential bottleneck is created by putting one person in charge of mitigating hundreds to thousands of clashes
- To cope with the volume of clashes, test strategies are employed to only look for clashes between certain sets of priority objects
- Leaving many others untested with the likelihood of slipping through the cracks

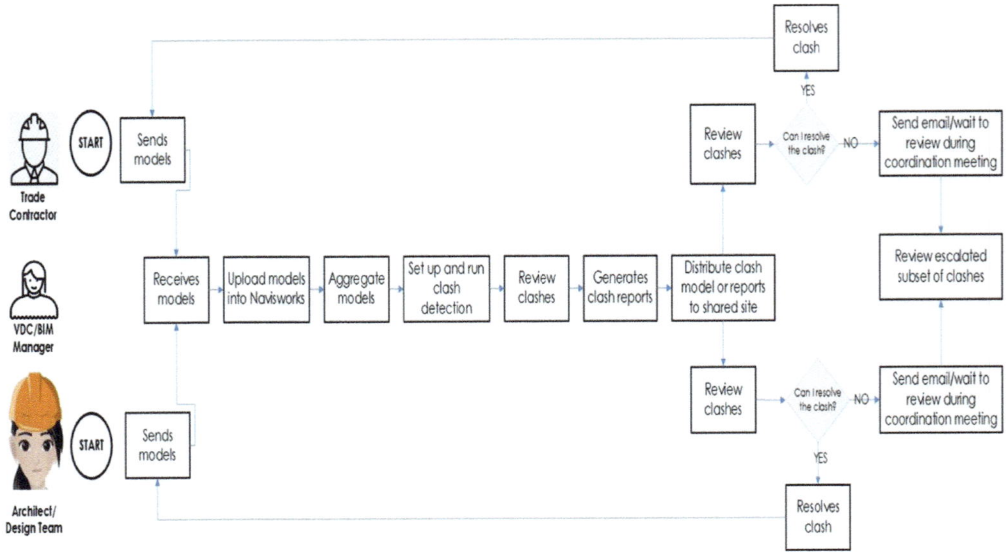

Figure 6.1 Traditional workflow of model coordination

Siloed information

- Everyone working through the coordination process has their own systems in place for file management and communications
- This only slows down file sharing and is sometimes disconnected

Time is money

- The coordination process is lengthy, and with project schedules becoming tighter, coordination time often bears the brunt of this tightening and is essentially shortened as well
- Every time a new model iteration is available, the VDC/BIM Manager has to run through the entire workflow causing them to spend a majority of their time on non-optimal activities.

Need for specialized software

- Traditional coordination processes usually require expensive software with expert users
- This can be an issue when stakeholders want to look at clashes on their own
- They need to have access to both the specialized software the VDC/BIM Manager is using as well as all of the multidiscipline models

Trade partner isolation

- More often than not, between coordination meetings with the group, trade partners are left to work on their own
- They adjust and update their given scopes of work and communication between the trade partners during this model adjustment time to minuscule.

6.5.4.2 Coordination Workflow in Autodesk Construction Cloud

To facilitate understanding a BIM-based workflow is presented in Figure 6.2.

Step 1: Create coordination spaces

- This is a folder in Autodesk Construction Cloud where project models to be clashed are stored
- This folder is created in BIM 360 Doc
- Also set permission levels (Figure 6.3)

Step 2: Set/Name coordination spaces

- Project Admin>Model Coordination>Choose created folder in Step 1 and give it a name (Figure 6.4)

Step 3: Drop and View 3D models

Figure 6.5 illustrates the dropping and viewing of 3D models in the coordination spaces.

Step 4: Automatic Clash Detection

Figure 6.6 illustrate automatic clash detection in BIM 360.

Note that AutoCAD 3D files, IFC, DWG, and RVT are the only supported files for coordination.

6.5.5 BIM 360 Project Management

This module facilitates project planning, document management, and workflow optimization. It includes tools for RFIs, submittals, meeting minutes, and daily reports, aiming to streamline communication and project tracking to keep teams aligned and projects on schedule.

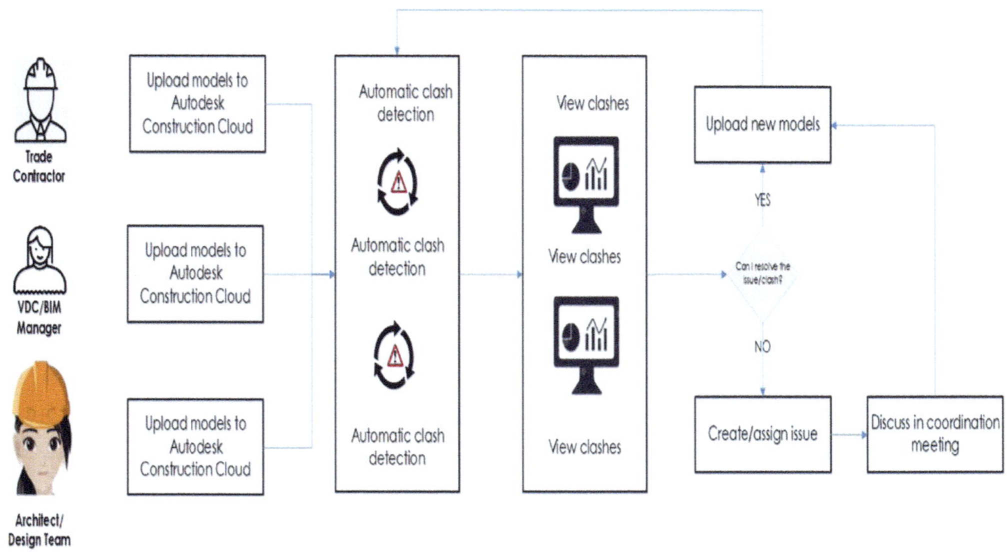

Figure 6.2 BIM-based workflow of model coordination

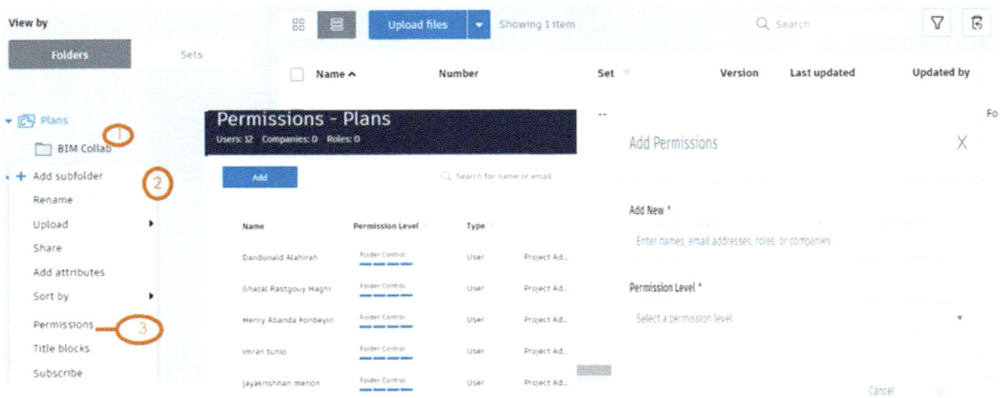

Figure 6.3 Creation of coordination spaces workflow

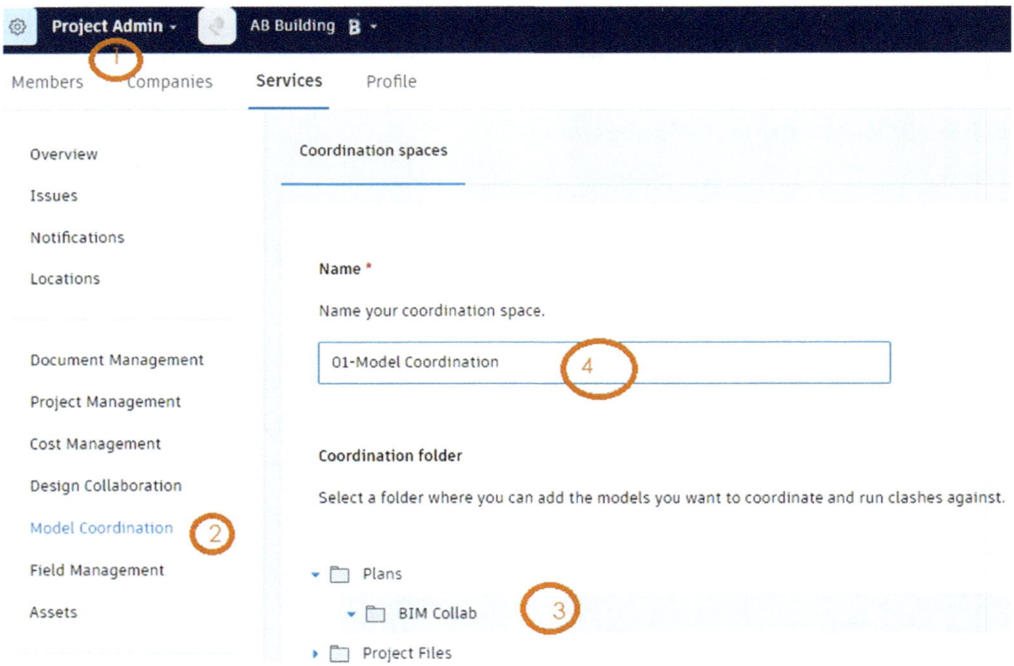

Figure 6.4 Setting and naming of the coordination spaces

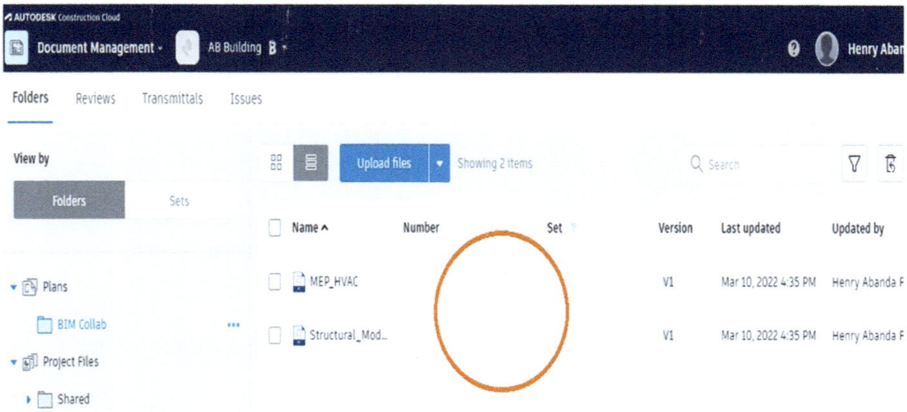

Figure 6.5 Dropping and viewing of 3D models in the coordination spaces

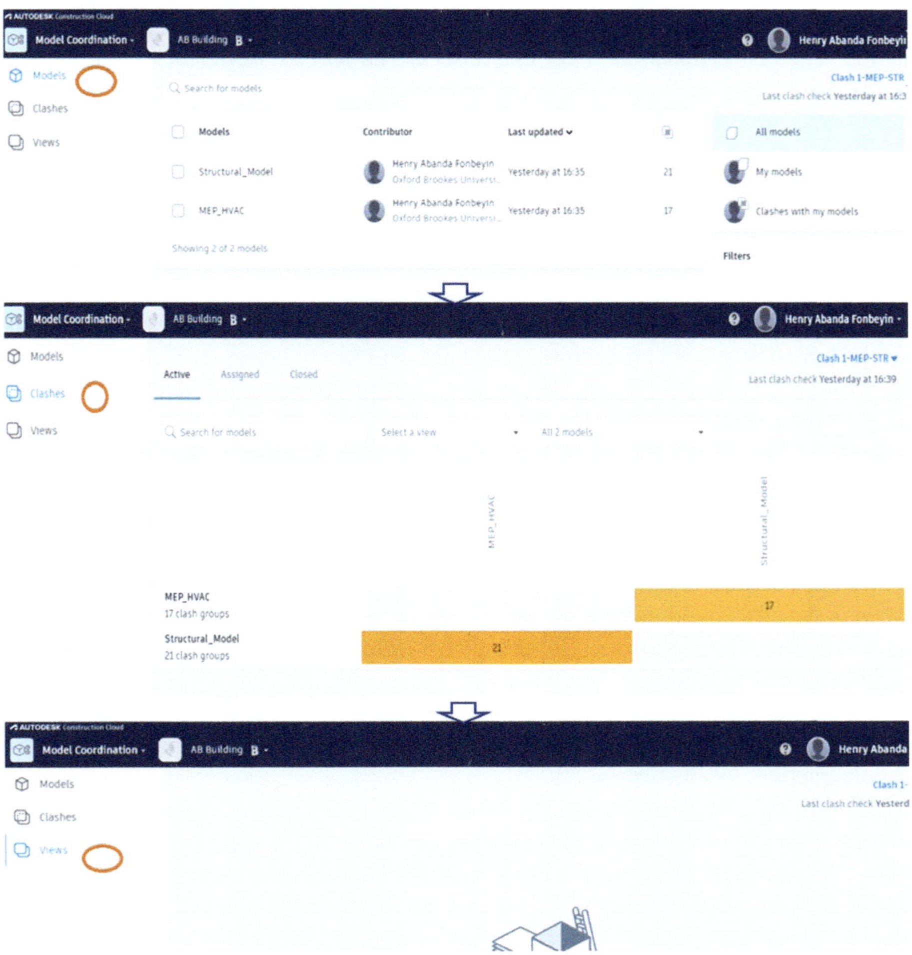

Figure 6.6 Automatic clash detection in BIM 360

Typical Information Types in Project Management

- RFI
- Submittals

6.5.5.1 Request for Information (RFI)

In construction, a RFI seeks the clarification of plans, drawings, specifications, and agreements. The *construction RFI* is a formal written process in which parties, such as the contractor and designer, clarify information gaps in construction documents.

- While this process sounds straightforward, RFIs often become a source of delay, expense, conflict, and even legal claims.
- Construction RFIs come into play at various points in the construction process. During bidding, a contractor may submit an RFI to gain a fuller understanding of the project's details and fine-tune a quote.
- More commonly however, a contractor or subcontractor will submit a construction RFI while building is under way in order to ask the designer or engineer to clarify a drawing or specification. The contractor or supplier may also use an RFI to document a concern about the specified materials, seek more information about the owner's intended use, or recommend a change or substitution of components.

Types of RFI

- **Design Coordination:** These RFIs involve organizing, communicating, and coordinating the design and associated documents among project participants.
- **Construction Coordination:** This category of RFI covers requests to organize and coordinate construction-related procedures, schedules, and safety items.
- **Constructability Issues:** These RFIs arise when the construction document articulates a detail or design that the contractor deems infeasible.
- **Design Change:** These RFIs seek to revise the design in order to facilitate construction or fix a construction mistake.
- **Design Clarification:** With this kind of RFI, the requester seeks more information about the design in order to better understand its details and how they relate to the project.
- **Change in Method or Process:** This category of RFI arises when the contractor wants to change a specified building or installation technique.
- **Deleted Scope:** These RFIs seek to remove work from the project.
- **Incomplete Plans or Specs:** This type of RFI points out errors or omissions in the plans and specifications.
- **Material Change:** With this type of RFI, the requester seeks to use a different material than the construction documents specify.
- **Site Conditions:** These RFIs emerge when builders discover a discrepancy between the description of site conditions in the construction documents and the actual site conditions.
- **Utility Conflict:** Use this type of RFI when utility lines (such as water pipes or power wires) stand in the way of proceeding with construction as originally planned.
- **Value Engineering:** This category of RFI seeks to make changes for the purpose of reducing cost, improving efficiency, or enhancing quality.

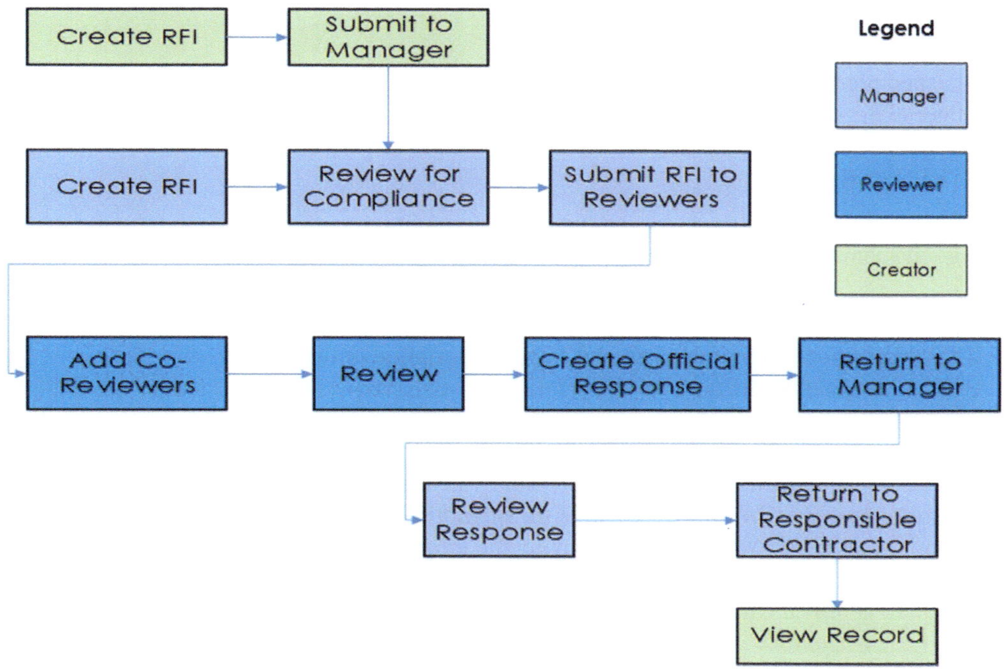

Figure 6.7 RFI workflow

- **Other Issues:** This category covers any other types of RFI that might come up, such as requests related to warranties, certifications and inspections, penalties, and non-design documents.
- **Timing and Phasing:** RFIs of this type occur when the builder needs to change the sequence of construction activities due to resource or labour constraints.

The RFI Workflow is presented in Figure 6.7.

6.5.5.2 Submittals

Construction submittals are defined by bizfluent as 'documents submitted by the contractor to the architect for his approval for use in a project,' while Lexology explains that 'Submittals consist of information provided by the contractor to the design professional for approval of equipment, materials, etc. before they are fabricated and delivered to the project.'

Before construction begins, every piece of equipment, material types, and even details such as the exact colour of paint need to be reviewed and approved via submittals. Depending on the individual project, construction submittals can involve thousands of different items.

These include:

- Product cut sheets that identify the manufacturer, specifications, and model number
- Shop drawings that lay out the dimensions of such prefabricated products such as trusses, cast concrete, windows, appliances, millwork, and more

- Colour and finish selections
- Colour charts
- Finished product components
- Material data
- Samples
- … and more

A good submittal log should at a minimum include the following:

- Specification section number/name/subsection: Where the requirement came from
- Title/description: Submittal name and a short description of the request
- Submittal type: What type of information is being requested (e.g., shop drawings vs. samples)
- Priority: Request/approval urgency
- Responsible contractor: Who will be providing the information
- Submittal manager: Who is responsible for submitting the item for review
- Submittal reviewer: Who will be approving the information
- Required date: When the submittal is due from the responsible contractor
- Required approval date: When the reviewer must approve the submittal by

Submittals in construction management are shop drawings, material data, samples, and product data. Submittals are required primarily for the architect and engineer to verify that the correct products will be installed on the project.

Shop drawings (or workshop drawings) might be prepared by contractors, subcontractors, suppliers, manufacturers, or fabricators. They generally relate to prefabricated components, showing how they should be manufactured or installed. They take design intent drawings and specifications prepared by the project design team and develop them to show in detail how the component will actually be manufactured, fabricated, assembled, or installed.

Commonly, shop drawings are prepared for a wide array of components, including but not limited to, structural steelwork, reinforcement, elevators, building services equipment, appliances, ductwork, piping, plumbing fixtures, windows, cabinetry, electrical and data layouts, fire protection systems, and more. These documents serve as a bridge between concept and reality, ensuring that the vision for the project is accurately translated into the physical world.

Submittals are documents created for consumption by other companies.

Submittals are broken up into two groupings:

- Items: Consist of a document or group of documents.
- Packages: Consist of a group of Items.

Packages are not required to go through the review process. Packages can be used to break up items that require revisions. For example, you could make items for electrical and mechanical systems. Once the related documents are in each of the items, a package can be sent for review with both items. This allows one of the items to be accepted while the other needs a revision (Figure 6.8).

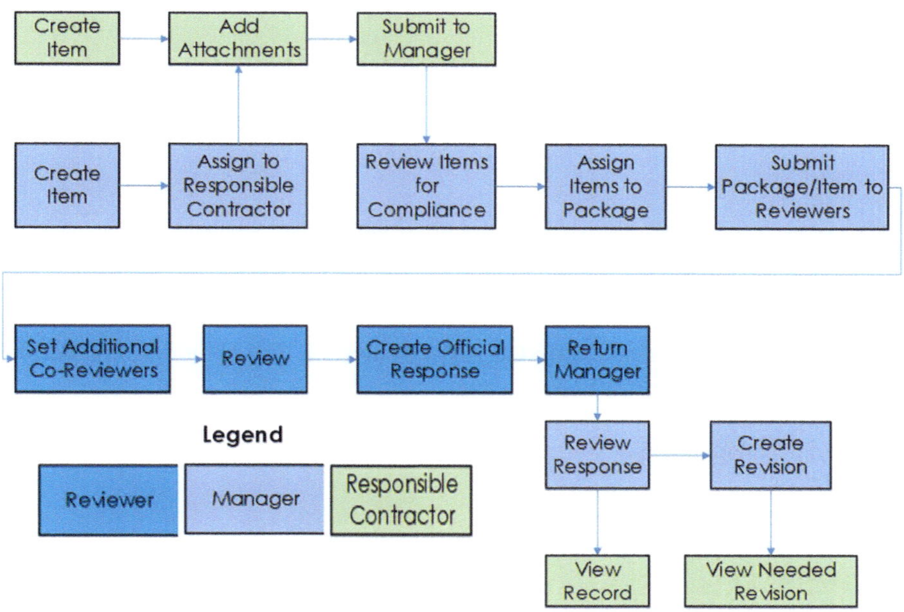

Figure 6.8 Submittals workflow

Responsible Contractor

- The primary subcontractor responsible for submitting the required shop drawings, material data, samples, and product data for approval.
  - Creating submittal items (if necessary).
  - Adding required material to submittal items for review.

6.5.5.3 Manager

Typically, the General Contractor. The Manager's primary abilities include:

- Creating items and assigning them to the responsible contractor.
- Assembling items into packages.
- Assigning packages to the reviewer and co-reviewers.

Managers can submit single items for review. In this case, a package is automatically created with the same name as the item.

6.5.5.4 Reviewer

The primary reviewer (typically, the Architect) is responsible for reviewing and approving the submitted materials. The Reviewer's primary abilities include:

- Reviewing the submittal.
- Assigning co-reviewers and managing official responses.

Co-Reviewers

- Engineers and consultants asked for review/feedback as part of the review process (usually by the Reviewer).

Distribution List

- Various stakeholders to be informed (notice only) upon status changes.

6.5.5.5 Meetings

The Meeting Minutes toolset can support any type of meeting. The Meetings Minutes toolset is found in the meetings tab of Project Management and allows you to:

- Create and delete meetings
- Track attendance
- Create agendas
- Create minutes – keep notes about the important decisions made in project meetings (Figure 6.9)

You can assign action items with due dates to attendees. Files can be attached to an action item for fast access during the conversation. RFIs and issues can be attached to meeting items for quick access during a meeting. You can create follow-up meetings for ongoing work.

6.5.6 BIM 360 Asset Management

This component of BIM 360 focuses on tracking and managing the lifecycle of construction assets from design through handover. It helps teams monitor the status, location, and performance

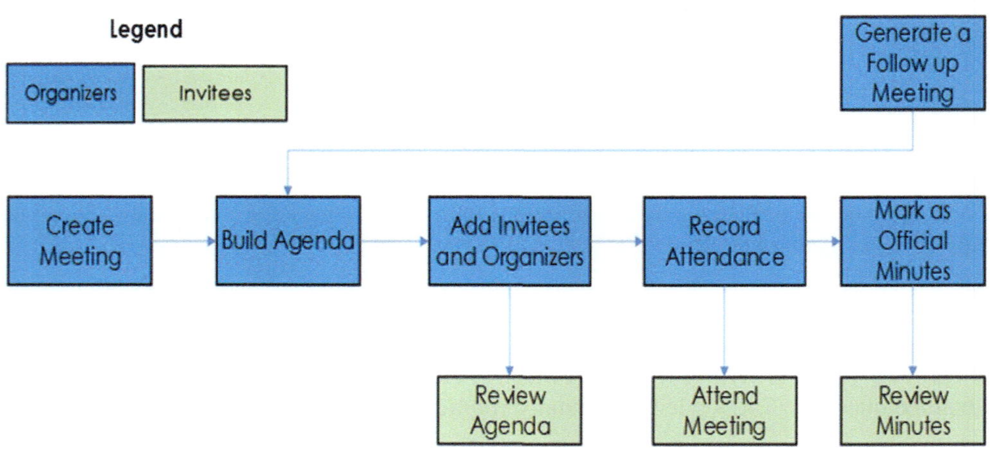

Figure 6.9 Meetings workflow

of materials, equipment, and other assets, ensuring they are delivered, installed, and maintained according to project specifications and schedules.

- the O&M cost amounts to 50–70% of the total annual facility operating costs (Rondeau et al., 2012)
- 85% of the entire lifecycle costs in facilities management is O&M (Teicholz, 2013).
- UK needs to retrofit 26 million homes by 2050 to reduce greenhouse gas emissions (The House, 2018)
- About 80% of the homes, we will be occupying in 2050 already exist (The House, 2018).

6.5.6.1 What Is an Asset?

It is an item, thing, or entity that has potential or actual value to an *organization*.

- An asset could be an item of building plant such as a boiler or chiller costing several thousand pounds; commercial kitchen equipment or a smaller item such as an office chair or flat screen display.

Value can be tangible or intangible, financial or non-financial, and includes consideration of *risks* and liabilities. It can be positive or negative at different stages of the *asset life*.

Physical assets usually refer to equipment, inventory, and properties owned by the organization.

Physical assets are the opposite of intangible assets, which are non-physical assets such as leases, brands, digital assets, use rights, licences, intellectual property rights, reputation, or agreements.

- Physical asset is simply known as 'asset'
- Virtual asset is known as 'information model'

The period from the creation of an asset to the end of its life is the asset life. An asset can provide potential or actual value to one or more organizations over its asset life, and the value of the asset to an organization can change over its asset life (Figure 6.10).

6.5.6.2 Categories (Types) of Asset

6.5.6.2.1 PROPERTY ASSETS

Definition: Property assets refer to real estate or land and buildings that a company or individual owns. These assets are typically immovable and form part of the real property portfolio. They are used for operations, investment, or other long-term purposes.

Examples:

- An office building owned by a company.
- A residential apartment complex held by a real estate investor.
- A commercial shopping mall owned by a retail chain.
- A school campus owned by an educational institution.

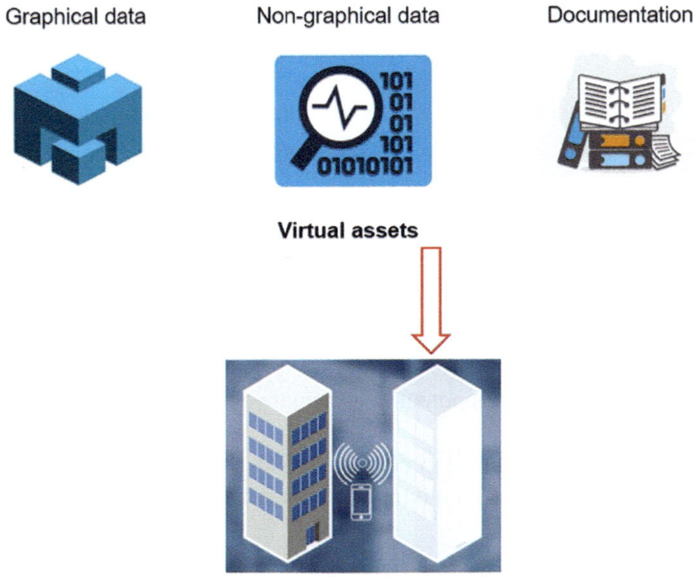

Figure 6.10 Physical versus Virtual (Digital) assets

6.5.6.2.2 FIXED ASSETS

Definition: Fixed assets are tangible, long-term assets that a company owns and uses in its operations to generate revenue. These assets are not intended for sale in the ordinary course of business and are generally immovable or integral to the building structure.

Examples:

- HVAC (heating, ventilation, and air conditioning) systems installed in a building.
- Elevators and escalators in a commercial building.
- Structural elements like walls, roofs, and flooring.
- Lighting systems and plumbing infrastructure within a building.

6.5.6.2.3 REMOVABLE ASSETS

Definition: Removable assets are items within a building that are not permanently affixed to the structure and can be detached and moved without causing significant damage. These assets are often part of the interior fit-out and are easier to replace or transfer.

Examples:

- Office furniture (e.g., desks, chairs, and cabinets).
- Appliances like refrigerators, washing machines, or microwaves.
- Modular partitions used in office spaces.
- Artwork and decorative items hung on walls.

6.5.6.3 *Asset Lifecycle*

Phase 1 of an Asset's Life – Planning & Acquisition

The initial cost (the capital outlay) of purchasing an asset only represents a small proportion of its total cost of ownership (TCO). The operating and maintenance costs typically outweigh the initial investment many times over. In its procurement guidance to whole-life costing, the UK's HM Treasury describes the acquisition cost as the tip of the iceberg, with the majority of the cost involved in the asset's operational phase including staff costs, training, maintenance, operation, withdrawal from service, depreciation, disposal, renewal, and rehabilitation.

Phase 2 of an Asset's Life – Operation

The operational phase of an asset's life is its longest and most useful period – it is when the asset is managed and used to deliver services to support the core business. Proactive asset management has a key role to play in this area.

Phase 3 of an Asset's Life – Maintenance

Proactive maintenance is at the very heart of asset management. A well-thought-through maintenance plan will improve the condition of, and extend the life of, the asset. Conversely, poor maintenance can reduce the useful life of an asset. Although an asset may be designed with a life of 20 years, good maintenance may mean that it could perform well beyond that period.

Phase 4 of an Asset's Life – Disposal

However good the asset management and maintenance planning in an organization, the time comes when an asset reaches the end of its useful life and it needs to either be recycled, sent to landfill, sold on, or donated (to an employee or third party) or dismantled, and the parts reused, scrapped, or sold.

6.5.6.4 *Asset Management*

It is the coordinated activity of an *organization* to realize value from *assets.*

The term 'activity' has a broad meaning and can include, for example, the approach, the planning, the plans, and their implementation.

The fundamental principle of asset management is to intervene at strategic points in an asset's normal life with optimized repair and maintenance activities, in order to maintain the performance of an asset and extend its life.

Asset Register: There is no fixed format or template for maintaining fixed asset register. The extent of details depends on the requirement of the management.

6.5.7 *Technology for Asset Management*

Asset management module is an asset and equipment tracking tool.

- It reduces the time it takes to access asset data in the field, perform commissioning, and resolve defects all while reducing risk
  - Centralized place to track and manage the lifecycle of project assets from design to handover
  - All asset-related documentation, including product information, cut sheets, and warranty documents, is stored in a central repository and connected to checklists and issues

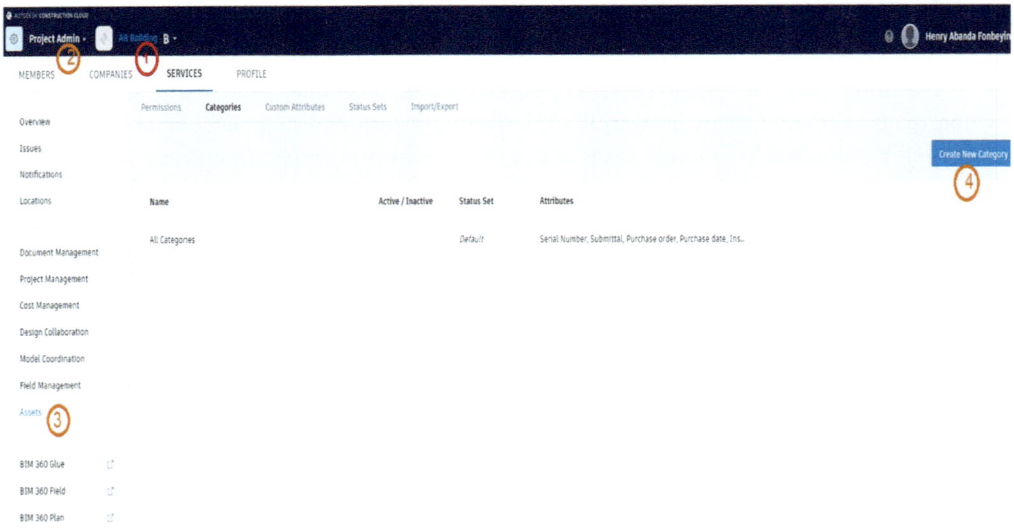

Figure 6.11 Creating Attributes

Setting up Assets

- Customization
  - Set up all custom categories, sub-categories, statuses, and attributes before adding in any assets
  - Within the 'Services' tab, select 'Assets' on the left side panel
  - You can customize:
    - Custom Categories
    - Custom Attributes
    - Status Sets

6.5.7.1 Creating a New Category

Figure 6.11 illustrates the creation of assets' attributes in BIM 360.

6.6 Conclusions

Given the detailed exploration of cloud-based BIM and the implementation of CDEs in the construction sector, as outlined in the preceding chapters, this conclusion aims to encapsulate the critical insights and forward-looking perspectives that pave the way to the subsequent focus on Information Management according to ISO 19650.

The journey through Cloud BIM and its practical applications via platforms like BIM 360 has highlighted the transformative potential of cloud-based solutions in enhancing project management, design, and execution across the construction industry. The adoption of Mobile BIM, leveraging the agility and flexibility of mobile technologies, further underscores the industry's move towards more accessible, real-time collaboration and information management.

The pivotal role of CDEs in underpinning effective digital collaboration cannot be overstated. By providing a centralized, accessible, and secure repository for project data, CDEs facilitate a seamless exchange of information, ensuring that all stakeholders are aligned and informed throughout the project lifecycle. This integration of information management practices is crucial for mitigating risks, reducing errors, and enhancing the efficiency and productivity of construction projects.

The exploration of BIM 360 tools – Document Management, Field Management, Design Collaboration, Model Coordination, and Asset Management – provides a comprehensive understanding of how cloud-based platforms can support various facets of construction management, from document control and issue resolution to real-time design collaboration and asset lifecycle management. These tools exemplify the practical benefits of leveraging Cloud BIM and CDEs, including improved coordination, reduced conflicts, and streamlined workflows.

However, the transition to digitalized construction practices is not without challenges. The reliance on specialized software, potential bottlenecks in model coordination, and the need for rigorous data security measures are among the hurdles that the industry faces. The evolution of project management approaches, as facilitated by BIM 360 Project Management, illustrates the industry's response to these challenges through innovative solutions that enhance communication, project tracking, and compliance with industry standards.

As the construction industry continues to evolve, the integration of Cloud BIM and CDEs will be instrumental in driving forward industry standards and practices. The next chapter's focus on Information Management according to ISO 19650 builds on this foundation, emphasizing the importance of structured information management processes in achieving more efficient, transparent, and collaborative construction projects. ISO 19650 represents a global standard for managing information over the whole life cycle of a built asset, highlighting the strategic significance of information management in realizing the full potential of BIM and digital construction technologies.

In conclusion, the exploration of Cloud BIM and CDEs has laid a solid groundwork for advancing digital collaboration and information management within the construction industry. The transition towards these technologies is not merely a trend but a fundamental shift towards a more integrated, efficient, and sustainable construction sector. The forthcoming discussion on Information Management according to ISO 19650 will further elaborate on how standardized information management practices can enhance the value and effectiveness of digital construction, ensuring that the industry is poised to meet the challenges and opportunities of the digital age.

Chapter 7

Information Management

7.1 Introduction

Over the past decade, there has been a significant surge of interest in digital construction technologies, particularly in building information modelling (BIM). However, this heightened interest has not fully translated into widespread adoption and practical implementation. Several challenges have hindered the uptake of these technologies in the industry. In response to these challenges, a suite of standards, including ISO 19650, was developed in 2018 to facilitate the implementation of BIM. Despite the introduction of the ISO 19650 standards, their intended objectives have yet to be fully realized (Ajayi et al., 2023; Ashworth et al., 2023; Malla et al., 2024). Many firms either fail to implement these standards or struggle with efficiency in their application (Malla et al., 2024).

The complexity of BIM is mirrored in the ISO 19650 standards, which are highly technical and consist of several sections requiring expert knowledge for proper interpretation. The aim of this chapter is to critically examine the relationship between information management and BIM compliance in projects. By exploring the intricacies of these standards and their practical implications, this chapter seeks to provide a deeper understanding of the challenges and opportunities in achieving effective BIM implementation.

7.2 Information Management

In practice, information is generated and/or consumed in the development of a project. This consumption and generation of information occurs in the different lifecycle phases of construction projects. It is therefore imperative to examine the management of information in the lifecycle of projects. This examination will be through the lens of ISO 19650.

Information management is the management and production of information during the life cycle of built assets using BIM. BS EN ISO 19650-1:2018 and BS EN ISO 19650-2:2018 support the information management process.

BS EN ISO 19650-1:2018 sets out the recommended concepts and principles for business processes across the built environment sector in support of the management and production of information during the life cycle of built assets.

The specific requirements for information management during the delivery of built assets are provided in BS EN ISO 19650-2.

DOI: 10.1201/9781003476740-7

7.3 Information Requirements

There are four key documents or resources which set out the appointing party's information requirements:

- Organizational information requirements (OIRs) – In relation to organizational objectives
- Project information requirements (PIRs) – In relation to the purpose, design and construction of an asset
- Asset information requirements (AIRs) – In relation to the operation and maintenance of an asset
- Exchange information requirements (EIRs) – in relation to an appointment

7.3.1 Organizational information requirements (OIRs)

OIRs explain the information needed to answer or inform high-level strategic objectives within the appointing party. These requirements can arise for a variety of reasons, including:

- strategic business operation
- strategic asset management
- portfolio planning
- regulatory duties
- policy-making

7.3.2 Project information requirements (PIRs)

PIRs explain the information needed to answer or inform high-level strategic objectives within the appointing party in relation to a particular built asset project.

7.3.3 Asset information requirements (AIRs)

AIRs set out managerial, commercial, and technical aspects of producing asset information. The managerial and commercial aspects should include the information standard and the production methods and procedures to be implemented by the delivery team.

The technical aspects of the AIR specify those detailed pieces of information needed to answer the asset-related OIR.

A set of AIR should be prepared in response to each trigger event during asset operation and where appropriate should also refer to security requirements.

Trigger event: planned or unplanned event that changes an *asset* or its status during its *life cycle* which results in *information exchange*

An example of a trigger event: planned or reactive maintenance, fire equipment inspection, component replacement, or change of asset management provider

7.3.4 Exchange information requirements (EIRs)

Employer's requirements provide a description of the client's requirements, including the specification for the building, the scope of services required from the contractor, and an allocation of risk for unknown items. Contractor proposals are then prepared in response to the employer's requirements. These present the contractor's suggested approach for designing and constructing the building, along with their price.

EIR is a 'pre-tender document setting out the information to be delivered, and the standards and processes to be adopted by the supplier/lead appointed party as part of the project delivery process.'

The EIR should be provided and sit alongside ER. While the project ER defines the nature of the built asset that the employer wishes to procure, the EIR should define the information about the built asset that the employer wishes to procure to ensure that the design is developed in accordance with the needs and that they are able to operate the completed development effectively and efficiently.

7.4 Information Models

7.4.1 *Asset Information Model (AIM)*

A model that compiles the data and information necessary to support asset management.

An AIM can provide graphical and non-graphical data and information as well as documents and metadata.

The information required for an AIM is defined in the AIR. This should be developed from OIR.

7.4.2 *Project Information Model (PIM)*

The PIM supports the delivery of the project and contributes to the AIM to support asset management activities. The PIM should also be stored to provide a long-term archive of the project and for auditing purposes.

For example, the PIM can contain details of project geometry, location of equipment, performance requirements during project design, method of construction, scheduling, costing and details of installed systems, components, and equipment, including maintenance requirements, during project construction.

Information model relating to the *delivery phase*

During the project, the PIM can be used to convey the design intent (sometimes called the design intent model) or the virtual representation of the asset to be constructed (sometimes called the virtual construction model).

Information Container: The term 'information container' refers to '...the unique identification of a file. For a construction project, an information container could be a cost plan, a programme, a drawing, a geometric model, etc.... Each information container should be identified through a unique ID.'

Data Drops: To ensure projects are properly validated and controlled as they develop, data are extracted from the evolving building information model and submitted to the client (appointing party) at key milestones. This submission of data is described as a 'data drop' or 'information exchange.'

Generally, data drops are aligned to project stages, and the information required reflects the level of development that the project should have reached by that stage. This might be considered analogous a stage report on a conventional project.

In other words:

- Data are produced and collated progressively as the project evolves
- The evolution of information comes in the form of documents, graphical, and non-graphical data

Data drops are likely to include:

- Models (Industry Foundation Class (IFC) models and native project information models).
- Data structures (such as COBIe files and schedules).
- Reports (typically PDFs, although native files can be more useable)

7.5 BIM Execution Plans (BEPs)

The BEP is a plan that supports information delivery during project execution. The BEP shall list the agreed targets for responsibility, timely delivery, exchange, reuse, and final handover to the clients. It will also list all of agreed elements as outlined in the Employers Information Requirements, the Brief, the BS1192:2007, PAS1192-2:2013, the CPIx Protocol, and the contract documents. This BEP is structured in accordance with PAS1192-2:2013.

There are two types of BIM EP – pre- and post-contract.

PAS 1192-2:2013 proposes that the pre-contract BEP is a direct response to the employer's information requirements (EIRs).

The post-contract-award BEP sets out how the information required in the employer's information requirements.

The main difference between pre-contract and post-contract BEP is that pre-contract addresses issues raised in the EIR and then with more detail post-contract award to explain the supplier's methodology for delivering the project using BIM.

The recent ISO 19650-1 uses appointment to mean contract, which is the agreed instruction for the provision of information concerning works, goods, and services. Based on this new terminology, the pre-contract and post-contract BIM Eps are pre-appointment and post-appointment BIM EPs.

7.6 Information Management Scope of Services

According to Clause 5.1.1 of ISO 19650-2, the appointing party should establish a scope of services where they wish to appoint a lead appointed party or third party to undertake all or part of the information management function.

Similarly, clause 5.3.1 requires a prospective lead appointed party to establish a scope of services where they wish to appoint a prospective appointed party or third party to undertake all or part of the information management function.

7.6.1 Record of Issue/Document Control

Definition: The Record of Issue or Document Control section of a BIM tendering document is a systematic process used to manage, organize, and track the various versions and revisions of documents within the project lifecycle. This section ensures that all stakeholders have access to the most current and accurate project information, mitigating the risks associated with working on outdated or incorrect documents. It includes details such as document version numbers, revision dates, authors or originators of the documents, and a log of changes or updates made to each document.

Examples:
A table listing all BIM model files, their version numbers, the date of the latest update, and a brief description of the changes made in the latest version.

A changelog document attached to the BIM project portal where each entry notes the document name, revision number, the date of revision, and a detailed description of the changes, who made the changes, and who approved the changes.

7.6.2 *Introduction*

Definition: The Introduction section of a BIM tendering document provides an overview of the project, including its scope, objectives, and the specific roles and responsibilities of the parties involved in the BIM process. This section sets the stage for the document by outlining the context of the project, the expected outcomes, and how the BIM methodology will be applied to achieve these outcomes. It may also include background information on the project, such as its significance, stakeholders, and any constraints or challenges anticipated.

Examples:
An opening statement that outlines the purpose of the BIM tendering document, the name and location of the project, and a brief description of the project's goals and expected benefits.

A summary of the project scope, including what will be designed and constructed using BIM technologies, the scale of the project, and any specific BIM objectives, such as improved coordination, enhanced design visualization, or lifecycle management.

Both sections are crucial for ensuring clarity, consistency, and efficiency in the BIM process from the tendering phase through to project completion and beyond. The Record of Issue/Document Control section ensures that everyone involved is working with the latest information, while the Introduction gives an overview of the project and sets expectations for its execution.

7.6.3 *Project Details (to Communicate Sufficient Understanding of the Project for the Third Party to Be Able to Submit a Tender Response)*

This section aims to provide all necessary information about the project to enable third parties to understand the scope, requirements, and specifications for submitting an informed tender response.

i **Key project information (including project name, project description)**

- Project Name: 'Downtown Commercial Complex'
- Project Description: A multi-use development comprising retail spaces, offices, and parking facilities, utilizing sustainable building practices and aiming for LEED Gold certification.
- Project Value: Estimated at $50 million.
- Procurement Route: Design-Bid-Build (if known), indicating that the project will first be fully designed by a hired architect, after which contractors will bid to undertake the construction.

ii **Programme (if known)**

- Timeline: The project is scheduled to start in Q2 2025 and is expected to be completed by Q4 2027.
- Milestones: Key milestones might include completion of design by Q3 2025, foundation work completion by Q1 2026, and so forth.

iii **Team (if known)**

- Architect: XYZ Architecture Firm, renowned for sustainable and innovative design solutions.
- Structural Engineer: ABC Engineering, known for their expertise in high-rise structures.
- MEP Consultant: MNO Technologies, specialists in energy-efficient building systems.
- This section could also include information on the project manager, quantity surveyor, and other key team members if known.

iv **Relevant technology solutions (if known)**

- BIM Software: Use of Autodesk Revit for 3D modelling and coordination among all disciplines.
- Collaboration Tools: Implementation of BIM 360 for cloud-based collaboration among stakeholders.
- Sustainability Analysis Tools: Use of Sefaira for early-stage energy and sustainability analysis to inform design decisions.

7.6.4 *Supporting Information*

This section provides additional documents, data, or information that supports the main tender submission. It can include references to existing models, drawings, reports, or any other material that helps in understanding the scope and requirements of the project.

Example:

- BEP from a previous project
- Sample BIM models or drawings relevant to the tender
- Reports on BIM standards and protocols used in past projects
- Case studies or references from similar projects

7.6.5 *Tasks*

This outlines the specific BIM-related tasks or deliverables that the bidder is expected to accomplish if awarded the project. This section is critical for setting the project's scope and expectations.

Example:

- Development of a BEP
- Creation and management of architectural, structural, and MEP models
- Coordination meetings and clash detection reports
- Maintenance of a project data environment
- Final model delivery and handover procedures

7.6.6 *Terms and Conditions*

This section details the legal, financial, and procedural terms and conditions under which the project will be executed. It includes payment terms, intellectual property rights, confidentiality agreements, and compliance with industry standards.

Example:

- Payment schedule and milestones
- Liability and insurance requirements
- Confidentiality agreements regarding project data
- Compliance with BIM Level 2 standards as a minimum requirement
- Software and version compatibility requirements

7.6.7 *Authority*

This specifies the organizational or individual authority overseeing the tendering process, decision-making, and project execution. It clarifies who has the power to make decisions, approve changes, and manage disputes.

Example:

- Designation of the project manager as the primary point of contact
- The role of a BIM coordinator or manager for model-related decisions
- Committees or boards responsible for review and approval stages

7.6.8 *Required Competency*

This section lists the skills, qualifications, and experience required from the bidding party to fulfil the project's BIM requirements effectively. It aims to ensure that the bidder has the necessary expertise to execute the project successfully.

Example:

- Certification in BIM management or relevant software proficiency
- Experience in similar scale projects
- Team composition, highlighting roles, and expertise in BIM processes
- Evidence of collaborative work in a multi-disciplinary BIM environment

7.6.9 *Probity Requirements (If Applicable)*

This part outlines the ethical and moral standards expected from the bidders, including transparency, accountability, and fairness in the tendering process. It may also include requirements for conflict-of-interest declarations and anti-corruption measures.

Example:

- Declaration of any potential conflicts of interest
- Statements affirming the absence of corrupt practices in the bidding process
- Requirements for third-party audits or integrity pacts

7.6.10 Tender Response Requirements

This encompasses all the detailed information and documentation that bidders must submit as part of their tender proposal for the BIM project.

i ***Fee proposal (against tasks)***
- A detailed pricing structure showing how much each task in the project scope will cost.

ii ***Fee drawdown (against programme)***
- Example: A payment schedule that outlines when payments are due, based on project milestones or specific phases of the work being completed.

iii ***Charge out rates (i.e., day rate(s)) and costs (including travel/expenses)***
- Example: A list of daily or hourly rates for each team member who will work on the project, along with any anticipated expenses, such as travel.

iv ***Capability and competency***
- Demonstration of capability (i.e., case studies including references)
  Example: Case studies of previous projects that are similar in scope and complexity, including references from those projects
- Demonstration of competency (i.e., CVs)
  Example: CVs of key team members showcasing their qualifications, experience, and roles in successful BIM projects

v ***Proposed delivery methodology***
Example: An outline of the approach and processes the bidder will use to deliver the BIM project, including software tools, collaboration practices, and quality control measures.

vi ***Assumptions***
Example: Statements about project-specific conditions or requirements that the bidder has assumed when preparing their proposal, such as site access or data availability.

vii ***Exclusions***
Example: A list of services or deliverables that are not included in the proposal and would be subject to additional charges if required.

viii ***Probity arrangements (if applicable)***
Example: Measures that will be taken to ensure fairness, transparency, and integrity in the bidding process, such as conflict of interest disclosures or third-party audits

ix ***Observations on supporting information provided (if applicable)***
Example: Comments or questions about the project brief or any documents provided with the tender request that may affect the proposal.

x ***Other supporting information relevant to the commission***
Example: Any additional documents or information that supports the bidder's proposal, such as certifications, policies, or innovative approaches to BIM projects.

7.6.11 Tender Return Procedures (Including Return Date, Person(s) to Return Tender)

This section specifies how and when the tender documents should be returned by the bidding parties. It includes details such as the submission deadline, the format in which the tenders should be submitted (electronic or hard copy), and any specific instructions for submission. It also identifies the person or department to whom the tender should be returned.

Example:

- The tender must be submitted in a sealed envelope marked 'Confidential – Tender Documents' to the Procurement Department by 5:00 PM on 15 March 2024. Electronic submissions should be sent to tenders@example.com in a single PDF file not exceeding 20MB.

7.6.12 *Tender Evaluation Criteria*

This part outlines the criteria against which the submitted tenders will be assessed. It can include a wide range of factors such as price, technical expertise, proposed methodology, compliance with BIM standards, previous experience in similar projects, and the quality of the BEP. The criteria may be weighted to indicate their relative importance in the evaluation process.

Example: Tenders will be evaluated based on the following criteria:

- Price (30%), BIM capability, and experience (25%),
- Quality of proposed methodology (20%),
- Compliance with project BIM standards (15%),
- and Sustainability approach (10%).

7.6.13 *Appointment (Proposed Date for Confirmation of Successful/Unsuccessful Tenderer and Feedback to Unsuccessful Tendering Organizations)*

This section communicates the timeline and process for notifying the successful and unsuccessful tenderers. It includes the proposed date by which the successful tenderer will be confirmed and outlines the procedure for providing feedback to those who were not selected.

Example:

- The successful tenderer will be notified by 10 April 2024 and will be expected to commence work on 1 May 2024. Unsuccessful tendering organizations will be informed by 15 April 2024 and can request feedback on their submission by scheduling a meeting with the procurement manager within two weeks of notification.

7.7 Conclusions

This chapter has traversed the landscape of information requirements, information models, BEPs, and the imperative delineation of information management scope of services. The essence of information management in the realm of BIM is encapsulated by its foundational pillars: information requirements and information models. Information requirements, comprising organizational, project, asset, and exchange information requirements, serve as the compass guiding all stakeholders through the lifecycle of construction projects. They ensure that every piece of data, from inception to operation and beyond, aligns with the overarching objectives of efficiency, sustainability, and functionality. Information models, including both the AIM and PIM, act as the blueprints for realizing these goals. They provide a multidimensional visualization of what is to be built, facilitating a level of precision and foresight previously unattainable.

The implementation of BEPs further cements the role of structured planning and execution in achieving project success. These plans are not merely documents but living frameworks that adapt and evolve, ensuring that the methodologies and technologies applied are always at the cutting edge of innovation. They encapsulate the who, what, when, and how of information exchange, ensuring clarity and coherence in the collaborative efforts of all parties involved.

Central to the execution of these plans is the information management scope of services. This crucial element delineates the responsibilities and expectations associated with managing the wealth of data BIM projects generate. By defining this scope, projects can harness the full potential of BIM, avoiding common pitfalls such as data silos, miscommunication, and inefficiencies. It ensures that the right information reaches the right people at the right time, facilitating informed decision-making and optimizing project outcomes.

As we prepare to delve into the benefits of digital construction technologies in the forthcoming chapter, the significance of the foundations laid in this chapter cannot be overstated. The digital transformation of the construction industry, heralded by the advent of BIM, is not just about adopting new tools but about a paradigm shift in how we conceptualize, plan, and execute construction projects. It represents a move towards a more collaborative, transparent, and efficient industry, capable of delivering projects that are not only financially viable but also environmentally sustainable and socially beneficial.

The journey through information management and BIM has laid the groundwork for appreciating the transformative potential of digital construction technologies. These technologies, from 3D printing to artificial intelligence and beyond, promise to revolutionize the construction industry. However, their successful integration and utilization hinge on a profound understanding of the principles and practices outlined in this chapter. As we advance, it is this foundational knowledge that will enable us to fully leverage digital technologies, pushing the boundaries of what is possible in construction.

Chapter 8

Benefits of Digital Construction Technologies

8.1 Introduction

Building on the insights from the previous chapter, where digital construction technologies were meticulously examined, we stand at a crucial juncture. The adoption of these technologies faces a formidable challenge, rooted in the lack of clear, widely recognized benefits. Without a deep understanding of the tangible and strategic advantages these technologies bring to the table, stakeholders may hesitate to invest time, resources, and trust in their implementation. This chapter aims to bridge that gap, offering a compelling argument for the adoption of digital construction technologies by illuminating their myriad benefits.

We will adopt a holistic approach to this exploration. Initially, we will delve into the general benefits of digital construction technologies, laying a foundational understanding of their impact on the construction industry at large. This includes improvements in efficiency, accuracy, and collaboration, among others. Following this, we will narrow our focus to the benefits of specific technologies, such as BIM, drones, and augmented reality (AR), to provide a more nuanced perspective on how each technology can be leveraged to achieve specific project goals.

To ensure a systematic and orderly approach to our discussion, the benefits will be aligned with the Royal Institute of British Architects (RIBA) Plan of Work. This alignment will not only structure our exploration but also contextualize the benefits within the well-established stages of a construction project, from conception through to completion and beyond. By correlating the advantages of digital construction technologies with the RIBA Plan of Work stages, we can offer a comprehensive and practical guide to how these technologies can enhance each phase of a project, ensuring stakeholders can clearly visualize the return on investment at every stage.

8.2 Benefits of Digital Construction Technologies

The benefits of digital construction technologies can significantly transform the construction industry, improving project outcomes, reducing costs, and enhancing overall operational efficiency. As digital technologies continue to evolve, their potential for disruption and positive impact on the construction sector becomes increasingly apparent. Some of the positive impacts of key advantages will be examined in the ensuing paragraphs.

DOI: 10.1201/9781003476740-8

8.2.1 *Benefits of BIM*

8.2.1.1 *Stage 0: Strategic Definition*

This initial stage focuses on understanding the client's business case and strategic brief. It's about defining the project's overall objectives, exploring the need for the building, and assessing the feasibility of the project. This stage sets the foundation for all subsequent work, ensuring that the project aligns with the client's strategic goals.

BIM facilitates early strategic decisions by providing a digital visualization of the project's potential, aiding stakeholders in understanding the scope and feasibility of the proposed development and their potential outcomes through digital models. This visualization aids in aligning project objectives and expectations among all parties involved. BIM's capability to integrate and manage extensive data sets from the outset supports comprehensive analysis and evaluation of design proposals, sustainability considerations, and cost implications, laying a solid foundation for informed decision-making and strategic planning (Kozlovska et al., 2023). This early integration of BIM processes ensures that project visions are both achievable and aligned with the stakeholders' objectives, optimizing project outcomes from the very beginning.

Another significant benefit of BIM in the Strategic Definition (Stage 0) of the RIBA Plan of Work is its ability to enhance stakeholder engagement and communication from the project's inception. By using BIM to create detailed and immersive 3D models, stakeholders can better understand the project's potential impacts, design intentions, and overall vision, facilitating more effective and inclusive decision-making processes. This early engagement helps to identify and address potential concerns, align expectations, and foster a collaborative project environment from the outset.

A third benefit of BIM in the Strategic Definition (Stage 0) of the RIBA Plan of Work is its ability to facilitate early risk identification and mitigation. Through BIM's comprehensive analysis and visualization capabilities, potential project risks and issues can be identified and addressed before they escalate into more significant problems. This proactive approach to risk management supports more effective planning, budgeting, and scheduling, ensuring a smoother project lifecycle and reducing the likelihood of unexpected challenges and costs (Smith, 2014).

8.2.1.2 *Stage 1: Preparation and Brief*

Building on the strategic definition, this stage involves developing the project brief in detail, conducting initial studies, and preparing for the concept design. It's about refining the project's objectives, considering the constraints, and outlining the requirements in a way that guides the design process.

BIM's role here is critical in refining project requirements and developing a detailed project brief. During the Preparation and Brief (Stage 1) of the RIBA Plan of Work 2020, BIM presents several pivotal benefits. Firstly, BIM significantly enhances stakeholder engagement by providing immersive, detailed visualizations of proposed projects, facilitating a deeper understanding and involvement in the early stages of development. Teams can simulate different scenarios, ensuring the project brief is both comprehensive and achievable. This enhanced engagement aids in fostering consensus among all parties involved and ensures that decision-making is both informed and collaborative. Secondly, BIM enables more accurate project scoping, allowing teams to precisely define the project's scale and complexity. This precision is instrumental in developing a comprehensive and accurate brief that reflects the true scope of the project, leading to better planning and resource allocation (Pacheco et al., 2024). Lastly, BIM's robust data management capabilities streamline the process of capturing and documenting project requirements. By efficiently organizing and presenting project data, BIM ensures that all stakeholder needs are accurately reflected in the project objectives, thereby aligning expectations and project outcomes. These benefits collectively underscore BIM's transformative potential in optimizing the preparation and planning phases of construction projects.

8.2.1.3 Stage 2: Concept Design

At this stage, the design team develops the conceptual approach to the project. This involves creating initial design proposals that reflect the project brief and exploring the project's architectural, building services, and structural systems. The concept design stage is crucial for setting the direction of the project.

Within the Concept Design stage (Stage 2) of the RIBA Plan of Work 2020, BIM presents transformative benefits, significantly enhancing the design process. Firstly, BIM's advanced visualization and simulation capabilities allow for the creation of detailed 3D models.

The ability to create accurate 3D models through BIM facilitates a comprehensive visualization of the project, allowing stakeholders to understand and evaluate the design intent effectively. This stage benefits significantly from the improved visualization and simulation capabilities provided by BIM, as documented by Abanda et al. (2014), enabling early detection of potential issues and fostering informed design decisions. Improved visualization accuracy through BIM enables architects and designers to create more realistic 3D visualizations, facilitating better decision-making and increased client satisfaction. Furthermore, 3D visualizations not only aid architects and designers in exploring spatial dynamics and aesthetic considerations early on but also support the evaluation of a building's functional performance. Such comprehensive visualization fosters a robust understanding of the project from its inception.

Moreover, BIM enhances collaboration among project stakeholders. By facilitating a shared digital workspace, it ensures real-time communication and iterative review of designs, enabling a collaborative approach that aligns visions and expectations across disciplines early in the project lifecycle (Azhar, 2011). This shared environment is crucial for integrating diverse expertise and feedback, leading to a more cohesive and informed design outcome.

Additionally, BIM aids in the early identification of potential design conflicts, streamlining the resolution process before these issues escalate into costly changes during construction. The integration of multi-disciplinary information into a unified model allows for the detection of clashes and discrepancies, promoting a proactive approach to design refinement (Sacks et al., 2022). This capability not only saves time and resources but also enhances the overall quality and feasibility of the design.

8.2.1.4 Stage 3: Spatial Coordination

Following the concept design, the spatial coordination stage involves the further development of the design, integrating technical systems and resolving any spatial issues. This stage ensures that the building systems are coordinated and fit within the architectural design, addressing both functional and technical requirements.

BIM applications play a pivotal role in Stage 3: Spatial Coordination of the RIBA Plan of Work 2020, offering substantial benefits that enhance both the efficiency and effectiveness of the construction process. Spatial coordination is a critical phase where the integration of architectural, structural, and building services systems is refined, ensuring that the conceptual design can be realized within the physical constraints of the project. BIM, as a collaborative and intelligent 3D model-based process, significantly contributes to this stage in several ways.

Firstly, BIM facilitates enhanced visualization and understanding of complex spaces (Liang and Sharudin, 2023). Through its 3D modelling capabilities, stakeholders can visualize the spatial relationships and constraints of the building systems more effectively than traditional 2D drawings allow. This improved visualization aids in identifying potential conflicts or issues early in the design process, enabling proactive resolution before construction begins, thereby reducing the likelihood of costly and time-consuming changes later on.

Secondly, BIM supports advanced clash detection and resolution. By integrating different disciplinary models into a single BIM environment, it becomes possible to automatically detect and visualize conflicts between architectural elements and building systems (Yönder and Çavka, 2024). This capability ensures that spatial coordination issues, such as ductwork clashing with structural elements, are identified and resolved during the design phase, significantly reducing the risk of on-site modifications.

Thirdly, BIM enhances collaboration among all project stakeholders. The centralized nature of a BIM model facilitates real-time sharing of information and updates, ensuring that all parties have access to the most current data (Raitviir and Lill, 2024). This collaborative approach streamlines the coordination process, as decisions and changes can be communicated and implemented efficiently, fostering a more integrated and cohesive design effort.

Furthermore, BIM applications contribute to a more sustainable design process by enabling the simulation and analysis of energy performance and other sustainability metrics during the spatial coordination stage (Abanda and Byers, 2016). By assessing the environmental impact of different design choices early on, teams can make informed decisions that enhance the building's sustainability without compromising on design integrity or increasing costs.

Lastly, the use of BIM during spatial coordination aids in the production of more accurate and detailed documentation. BIM models can automatically generate construction documents and drawings that reflect the coordinated design, ensuring that the information conveyed to construction teams is precise and comprehensive. This accuracy minimizes misunderstandings and errors during construction, contributing to a smoother project execution.

In conclusion, BIM applications offer significant advantages during Stage 3: Spatial Coordination of the RIBA Plan of Work 2020, including improved visualization, advanced clash detection, enhanced collaboration, support for sustainable design, and the production of accurate construction documents. By leveraging these benefits, project teams can navigate the complexities of spatial coordination more effectively, leading to better-designed, more sustainable, and efficiently executed projects.

8.2.1.5 Stage 4: Technical Design

In this stage, the design is developed in detail to enable construction. This includes finalizing the design and specifications of all elements and systems, ensuring compliance with regulatory requirements, and preparing for construction. The technical design stage is critical for translating the concept into a viable construction plan.

This stage, critical for translating conceptual designs into detailed plans for construction, benefits immensely from BIM's capabilities. Below, we explore these benefits in detail, highlighting how BIM applications revolutionize technical design processes.

Enhanced Collaboration and Coordination: BIM facilitates a highly collaborative environment among architects, engineers, and other stakeholders by providing a shared, digital representation of the building. This collaborative approach ensures that all parties have access to the most current design, reducing the risk of miscommunication and errors. BIM's ability to integrate data from different disciplines helps in identifying and resolving clashes between structural, mechanical, electrical, and plumbing (MEP) systems before construction begins, saving time and reducing costs.

Improved Accuracy and Quality of Documentation: BIM applications generate detailed digital models that contain precise information about the physical and functional characteristics of a building. This level of detail enhances the accuracy of technical drawings and documentation, ensuring that they accurately reflect the intended design and comply with relevant standards and regulations. The high-quality documentation produced through BIM supports better decision-making and contributes to the overall quality of the construction project.

Efficiency and Time Savings: By automating routine drafting tasks and enabling the reuse of model data for different purposes, BIM significantly reduces the time required to produce technical drawings and specifications. This efficiency not only accelerates the design phase but also allows more time for design optimization and innovation. Furthermore, BIM's ability to detect clashes and potential issues early in the design process reduces the need for rework, further saving time and resources (Azhar, 2011).

Cost Reduction: BIM's impact on reducing construction costs is well-documented. By improving the accuracy of cost estimations and facilitating the early detection of issues, BIM helps in avoiding unexpected expenses during construction. The detailed models produced through BIM enable more precise material take-offs and waste reduction, contributing to further cost savings.

Enhanced Sustainability: BIM applications support the design of more sustainable buildings by enabling the analysis of energy performance and other sustainability metrics during the technical design stage. This allows for informed decisions about materials, construction methods, and energy systems that can reduce the environmental impact of the building.

Risk Management: The detailed information provided by BIM models aids in identifying and managing risks associated with technical design and construction. By visualizing the building in detail before construction begins, potential structural issues and other risks can be addressed proactively, enhancing safety and reducing the likelihood of costly errors.

8.2.1.6 Stage 5: Manufacturing and Construction

This stage marks the transition from planning to physical construction. It involves manufacturing (if applicable), constructing the building according to the technical designs, and managing the construction process to ensure quality, safety, and adherence to the timeline and budget.

BIM transforms the construction phase by providing a digital blueprint that can be used for accurate construction planning, scheduling, and cost estimation, significantly reducing the risk of errors and misunderstandings on site.

BIM's facilitation of project scheduling and clash detection is integral to this phase (Kekana et al., 2015), improving coordination among trades and optimizing construction management through enhanced visualization and integrated project delivery approaches.

Project Management and Coordination: The application of BIM during the manufacturing and construction phase significantly enhances project management and coordination. Through BIM, project managers can visualize the entire construction process, identify potential conflicts before they occur, and optimize schedules and resources. This pre-emptive approach to conflict resolution is crucial for minimizing delays and avoiding cost overruns. BIM also supports advanced scheduling techniques like 4D BIM, integrating the project timeline with the 3D model to visualize the construction process over time, facilitating better planning and execution (Sacks et al., 2013).

Cost Reduction and Budget Management: BIM contributes to cost reduction and budget management through detailed cost estimation and control. By linking the model to cost databases, BIM enables the automatic generation of accurate quantity take-offs and cost estimates, which are vital for budgeting and financial management. This level of detail and accuracy in cost estimation helps in making informed decisions, managing finances more effectively, and ensuring that the project stays within budget.

Enhancing Sustainability: Sustainability is a critical concern in contemporary construction projects. BIM aids in the selection of materials and construction methods that minimize environmental impact and enhance energy efficiency (Atashbar and Noorzai, 2023). Through simulation tools integrated with BIM, project teams can analyse the energy performance of different designs and materials, enabling the selection of the most sustainable options. Moreover, BIM facilitates the optimization of material usage and waste reduction, contributing to more sustainable construction practices.

Quality Control: Quality control is paramount in the manufacturing and construction phase, and BIM significantly improves this aspect. The detailed models created in BIM provide a comprehensive reference for the construction team, ensuring that the work conforms to the specified standards and designs. BIM also supports the implementation of building performance analysis, which helps in ensuring that the completed facility will meet the expected performance criteria, including structural integrity, energy efficiency, and comfort.

Improving Communication Among Stakeholders: Effective communication among stakeholders is essential for the success of any construction project. BIM facilitates this by providing a shared platform for information exchange, where the model serves as the single source of truth (Khosrowshahi, 2017). This centralized approach to information management helps in reducing misunderstandings and miscommunications, ensuring that everyone involved in the project has access to the latest information. Moreover, the visualization capabilities of BIM allow non-technical stakeholders to understand the project details more clearly, enhancing collaboration and decision-making.

8.2.1.7 Stage 6: Handover and Close-Out

Once construction is complete, this stage focuses on finalizing the project, including completing any outstanding work, addressing any defects, and handing over the building to the client. It also involves compiling and delivering all project documentation necessary for the operation of the building. BIM streamlines the handover process by ensuring that all project data and documentation, including a digital twin of the building, are accurately compiled and easily accessible for future maintenance and management.

BIM in Enhancing Documentation and Information Exchange: BIM facilitates the comprehensive collection and organization of project data, which is crucial during the handover. The model acts as a central repository for all project-related information, including design details, materials specifications, and operational manuals, which can be seamlessly transferred to the client and facility managers. This ensures that the handover process is smooth and that the client receives all necessary documentation in an organized and accessible digital format.

The use of BIM also streamlines the creation of as-built documentation, providing a detailed and accurate representation of the building as constructed. This documentation is vital for future maintenance, renovation, or expansion projects, ensuring that the building's lifecycle is efficiently managed.

BIM's Role in Operational and Maintenance Planning: BIM extends beyond the construction phase, offering significant benefits during the building's operational phase. The detailed information stored in BIM models aids facility managers in planning and executing maintenance activities efficiently. For instance, BIM can be used to simulate the performance of building systems, identify potential maintenance issues before they occur, and plan maintenance activities to minimize disruption to building operations (Hauashdh et al., 2024).

Furthermore, BIM enables the integration of asset management systems (AMSs), where all asset information, including warranties, maintenance schedules, and manufacturer details, is stored within the BIM model. This integration ensures that facility managers have immediate access to up-to-date information, optimizing the building's long-term maintenance and operational efficiency.

Enhancing Client Understanding and Satisfaction: The visual and interactive nature of BIM models enhances client understanding and satisfaction during the handover process. Clients can explore the digital model, gaining a comprehensive understanding of the building's design, systems, and functionalities. This interactive experience facilitates a more in-depth appreciation of the building, enabling clients to make informed decisions about its use and maintenance.

BIM also supports the development of customized user manuals and operational guides based on the specific needs and preferences of the client, further enhancing their ability to manage and maintain the building effectively.

Sustainability and Lifecycle Management: Sustainability is a critical concern in modern construction, and BIM plays a vital role in ensuring sustainable building practices are considered throughout the building's lifecycle. During the handover and close-out stage, BIM can be used to provide detailed information on the building's environmental performance, including energy consumption patterns, water usage, and waste management strategies. This information enables clients to manage their buildings more sustainably, reducing environmental impact and operational costs.

Additionally, BIM facilitates the planning of future refurbishments or modifications in a manner that considers sustainability, allowing for the efficient reuse or recycling of materials and ensuring that any changes adhere to sustainable building practices.

8.2.1.8 Stage 7: Use

The final stage of the RIBA Plan of Work extends beyond the completion of construction, focusing on the building's operation, maintenance, and lifecycle performance. It includes tasks such as evaluating how the building meets its intended use, assessing performance, and making any necessary adjustments or improvements.

In the final stage, BIM continues to deliver benefits by enabling the efficient management and operation of the building. The detailed information contained within the BIM model supports facilities management and space planning and can contribute to sustainability and energy management initiatives. The role of BIM in promoting sustainability through energy efficiency analysis and material selection aligns with the operational phase of buildings (Al-Saeed and Ahmed, 2019), supporting performance monitoring and the integration of renewable energy sources for long-term sustainability.

Facilities Management: One of the primary applications of BIM during the Use stage is in facilities management. BIM provides detailed information about the building's physical assets, allowing facilities managers to access comprehensive data about each component, from HVAC systems to lighting fixtures. These data can be used to manage the building more effectively, ensuring that maintenance is carried out efficiently and on schedule. Moreover, BIM can facilitate the integration of Internet of Things (IoT) sensors and other smart technologies into facilities management, allowing for real-time monitoring and management of building systems.

Asset Management: BIM also plays a crucial role in asset management by enabling the tracking of asset performance over time. With BIM, asset managers can predict when a piece of equipment is likely to fail or require maintenance, allowing for proactive rather than reactive maintenance strategies. This not only extends the lifespan of assets but also reduces downtime and operational costs. Furthermore, BIM's data-rich environment supports the implementation of advanced AMSs, such as computer-aided facility management (CAFM) and integrated workplace management systems (IWMSs), enhancing the efficiency of asset management practices (Teicholz, 2013).

Space Management: Effective space management is essential for optimizing the use of space within buildings, especially in complex structures or facilities with dynamic space requirements. BIM facilitates space management by providing accurate, up-to-date information about space usage and occupancy patterns. This information can be used to make informed decisions about space allocation, adjustments, and improvements, ultimately enhancing the functionality and flexibility of the building space. BIM models can also be integrated with space management software to automate space planning and optimization processes.

Energy Analysis: BIM supports sustainable building operations through energy analysis, allowing building owners and operators to identify opportunities for energy savings and to monitor energy consumption patterns. By integrating BIM with energy simulation tools, stakeholders can assess the impact of various factors on energy use, such as changes in occupancy, equipment usage, and environmental conditions. This enables the implementation of energy-efficient strategies, such as retrofitting or adjusting HVAC operations, to reduce energy consumption and carbon emissions.

Maintenance Scheduling: Maintenance scheduling is another critical application of BIM during the Use stage. BIM provides detailed information on the maintenance requirements of building components, including manufacturer recommendations and warranty information. This information can be used to develop a comprehensive maintenance schedule that ensures all parts of the building are maintained appropriately, reducing the risk of unexpected failures and extending the lifespan of building components. BIM can also be integrated with maintenance management systems to automate the scheduling and tracking of maintenance activities.

Lifecycle Management: Finally, BIM facilitates lifecycle management by providing a platform for tracking the performance and condition of a building over its entire lifecycle. This includes monitoring changes in the building structure, systems, and occupancy, as well as managing renovations and upgrades. By maintaining a comprehensive and up-to-date BIM model throughout the building's lifecycle, stakeholders can make informed decisions about the building's future, ensuring that it continues to meet the needs of its users and operates efficiently and sustainably.

8.2.2 *Benefits of Robots in Construction in Project Delivery*

A key advantage of this robotic technology is its potential to significantly reduce injuries related to being struck by moving vehicles, which currently account for approximately 18% of all fatal injuries within industries. The technology minimizes the need for workers to guide construction traffic, thereby reducing their risk of being struck by moving vehicles. Moreover, the sensors on these vehicles are designed to respond to their surroundings, meaning that if a construction worker is in close proximity to danger, the robot can detect this and stop, preventing potential injuries or fatalities.

Another significant benefit of these robots is their adaptability. They can be easily deployed to different construction sites, adjusting seamlessly to new environments. With the use of Light Detection and Ranging (LiDAR) sensors and GPS, the robot can accurately perceive its surroundings and adapt to any changes on the site. GPS technology also ensures that the robot is aware of its final destination, regardless of the site's layout. Additionally, these electric-powered vehicles are highly energy-efficient, reducing carbon emissions by as much as 95% (Korane, 2017). This not only enhances energy efficiency but also supports the construction industry's move towards sustainability, which is crucial in addressing climate change.

The system offers a multitude of benefits, particularly in the area of health and safety. For instance, the technology significantly reduces the physical workload for masons, cutting heavy lifting tasks by at least 50% due to the robot's ability to lift bricks from the conveyor belt onto the wall. Some manufacturers claim that this reduction could be as high as 80% (Construction Robotics, 2019). Given the elevated incidence of musculoskeletal injuries in the industry – 1.1% higher than the industry average – this type of robot is a crucial step towards lowering injury rates by reducing lifting, bending, and the repetitive nature of bricklaying.

Increased productivity is another advantage, as the reduced reliance on human labour allows for more efficient project timelines. According to the robot developer, productivity in masonry could increase by three to five times, which would help mitigate the 25% of construction projects that are typically delayed, as mentioned in section one (Construction Robotics, 2019). The robot's ability to work continuously without breaks, requiring only one mason to operate, further enhances efficiency, allowing for longer work periods.

A similar study examined a placing robot's ability to identify discrepancies on site compared to the drawn model and implement changes based on on-site readings. Lundeen et al. (2019) found that while the final finish did not meet the robot's optimal standards, it was within acceptable limits and comparable to human workmanship. The robot demonstrated a positioning error of only 0.11mm, indicating a high level of precision and adaptability.

The widespread application of robots has led to significant improvements in productivity, increasing the annual growth rate of labour production by approximately 36% and contributing an estimated $1.3 trillion in value to the global economy each year (Graetz and Michaels, 2018).

8.2.2.1 Stage 0: Strategic Definition

The application of robotics, particularly collaborative robots (cobots) designed to work alongside humans in various capacities – in the Stage 0: Strategic Definition of the RIBA Plan of Work 2020, offers several compelling benefits. This stage, focused on understanding the client's business case and strategic brief, is foundational for the success of any construction project. Integrating cobots at this initial phase can significantly enhance the planning and decision-making processes.

Enhanced Feasibility Studies and Data Collection: Cobots can play a pivotal role in conducting detailed feasibility studies and collecting data essential for defining a project's strategic objectives. By employing drones or other robotic devices equipped with sensors and cameras, teams can gather comprehensive site data, including topography, existing conditions, and environmental factors, with greater accuracy and efficiency than traditional methods (Linner and Bock, 2012). These data can be instrumental in assessing the viability of a project, identifying potential challenges early, and making informed decisions.

Improved Stakeholder Engagement: The use of interactive and AR cobots can revolutionize stakeholder engagement during the strategic definition phase. By creating immersive visualizations of the proposed project, stakeholders can better understand the vision and objectives, facilitating more effective communication and collaboration. This technology can help bridge the gap between technical jargon and stakeholder understanding, leading to a more inclusive and informed decision-making process (Wang et al., 2015).

Risk Assessment and Mitigation: Cobots equipped with predictive analytics capabilities can analyse vast amounts of data to identify potential risks at the outset of a project. This proactive approach to risk management enables project teams to devise strategies to mitigate identified risks before they become issues, ensuring a smoother project lifecycle and potentially saving time and resources.

Robotic technology, particularly drones equipped with LiDAR, plays a pivotal role in the strategic definition of a project. These drones can quickly and accurately map out a site, providing essential data for feasibility studies and initial planning.

8.2.2.2 Stage 1: Preparation and Brief

During the Preparation and Brief stage, environmental impact assessments benefit from robotic technology. Drones can monitor wildlife and vegetation without disturbing the ecosystem, offering a non-invasive method to gather necessary data for environmental considerations in project planning.

8.2.2.3 Stage 2: Concept Design

In the Concept Design stage, the use of robotic arms for rapid prototyping allows for efficient exploration of design alternatives.

8.2.2.4 Stage 3: Spatial Coordination

This stage, crucial for integrating and coordinating spatial and technical aspects of building designs, benefits significantly from the precision, efficiency, and collaborative potential offered by cobots. The discussion below explores these benefits, supported by a framework of academic and industry insights.

Precision and Accuracy: Cobots are engineered to execute tasks with a level of precision that significantly surpasses human capabilities. In the context of spatial coordination, this precision is invaluable for ensuring that the integration of structural elements, mechanical systems, and architectural aesthetics adheres strictly to the specified dimensions and tolerances. The accuracy of cobots in measuring and placing components can minimize errors, reduce waste, and ensure the structural integrity of the building.

Efficiency and Productivity: The deployment of cobots can lead to substantial gains in efficiency and productivity. Unlike their human counterparts, cobots can work around the clock without fatigue, maintaining consistent output and speed. This capability is particularly beneficial in the fast-paced environment of construction projects, where delays can lead to significant financial repercussions. Furthermore, cobots can be programmed to perform repetitive tasks, freeing up human workers to focus on more complex, creative, or strategic aspects of the spatial coordination process (Lee and Kim, 2018).

Enhanced Collaboration: Cobots are designed to work alongside humans, complementing their skills rather than replacing them. This collaborative approach fosters a synergistic work environment where the strengths of both humans and robots are leveraged. For instance, humans can provide the creative and adaptive thinking necessary for overcoming unforeseen challenges, while cobots can execute the defined tasks with high precision and reliability. This collaboration can enhance the overall quality of the spatial coordination stage, ensuring a more seamless integration of various building systems.

Risk Reduction: Working in construction environments involves exposure to various hazards. By integrating cobots into spatial coordination tasks, the risk of injury to human workers can be significantly reduced. Cobots can be deployed in environments that are unsafe for humans, such as areas with toxic materials or extreme heights. This not only ensures the safety of the workforce but also helps in maintaining a consistent pace of work regardless of the environmental conditions.

8.2.2.5 Stage 4: Technical Design

The integration of robotic applications in the Technical Design stage of the RIBA Plan of Work 2020 presents a transformative opportunity for the construction industry. Robotics, with their precision, efficiency, and ability to handle complex tasks, can significantly enhance the design process, offering benefits that range from increased accuracy and efficiency to innovation in construction methods. This discussion explores the multifaceted advantages of robotic applications within this crucial phase of construction projects.

Increased Accuracy and Efficiency: Robotic technologies have been instrumental in elevating the accuracy of technical designs. Through automated processes, robots can execute tasks with a level of precision that significantly reduces the margin for error (Linner and Bock, 2012). For instance, robotic arms equipped with cutting-edge software can produce detailed models and components, ensuring that the specifications outlined during the Technical Design stage are met with unparalleled accuracy. This precision is critical in minimizing costly revisions and delays, thereby enhancing overall project efficiency.

Enhanced Innovation and Creativity: Robotic applications foster innovation in the Technical Design stage by enabling architects and engineers to explore complex geometries and innovative construction techniques that would be difficult, if not impossible, to achieve manually. Robots can fabricate components with intricate designs or materials that require precise manipulation, broadening the scope of what is possible in architectural design and construction.

Improved Safety and Reduced Labour Costs: The use of robotics in technical design also contributes to improved safety on construction sites. By automating the production of high-risk components, robots can reduce the need for human workers to engage in potentially hazardous tasks. This not only enhances safety but can also lead to significant savings in labour costs and insurance premiums, as the risk of accidents and associated liabilities decrease.

Sustainability and Waste Reduction: Robotic applications promote sustainability in the Technical Design phase through precise material usage and waste reduction. Robots can optimize the use of resources by accurately cutting and assembling materials, minimizing offcuts and excess. This efficiency is crucial for sustainable construction practices, as it reduces the environmental footprint of projects.

At the Technical Design stage, robotic simulations of construction processes can identify potential issues before on-site work begins, reducing the risk of costly errors. Such simulations have been shown to improve construction, ensuring that projects are completed on time and within budget.

8.2.2.6 Stage 5: Manufacturing and Construction

The integration of robotic technologies in the Manufacturing and Construction phase (Stage 5) of the RIBA Plan of Work 2020 has been heralded as a paradigm shift, bringing about profound efficiencies, quality improvements, and safety enhancements. This discussion explores the multifaceted benefits of robotic applications in these critical phases, drawing on academic and industry insights to underscore their transformative potential.

Efficiency and Productivity: Robotic technologies significantly enhance efficiency and productivity in construction and manufacturing processes. Robots, with their ability to work continuously without fatigue, offer unparalleled speed and precision (Linner and Bock, 2012). For instance, automated robotic arms can lay bricks or conduct welding tasks at a pace far exceeding human capabilities, thereby reducing project timelines and increasing throughput. Furthermore, the implementation of robotics in offsite manufacturing processes allows for the prefabrication of building components with minimal waste and higher precision, optimizing material usage and reducing the carbon footprint of construction activities.

Quality Enhancement: Robotic applications ensure a high degree of accuracy and consistency, directly translating to improved quality of construction. Unlike manual labour, which can vary in precision due to human error or fatigue, robots maintain consistent performance levels. This consistency is crucial in tasks that require high precision, such as the installation of MEP components, ensuring that the final construction meets stringent quality standards (Wang et al., 2018).

Safety Improvements: The adoption of robotics in construction and manufacturing significantly enhances on-site safety. Robots can be deployed in hazardous environments or assigned tasks that pose risks to human workers, such as working at heights or handling dangerous materials. By minimizing human exposure to such risks, robotic technologies contribute to a reduction in workplace accidents and associated costs.

Cost Reduction: While the initial investment in robotic technology may be significant, the long-term cost savings are substantial. Robotics reduce the need for manual labour, lower the incidence of errors and rework, and shorten project timelines. These factors collectively contribute to a lower overall project cost. Additionally, the precision afforded by robotic applications reduces material wastage, further contributing to cost efficiency.

Sustainability: Robotic applications in manufacturing and construction support sustainability goals by optimizing material usage and reducing waste. The precision of robotic systems ensures that materials are used efficiently, minimizing the environmental impact of construction projects. Moreover, the ability to work with a variety of materials encourages the use of sustainable and recycled materials, promoting eco-friendly construction practices (Ding et al., 2014).

In conclusion, the application of robotics in the Manufacturing and Construction phase of the RIBA Plan of Work 2020 offers compelling benefits, including enhanced efficiency, productivity, quality, safety, cost savings, and sustainability. As the construction industry continues to evolve, the integration of robotic technologies represents a critical step forward, promising a future where construction projects are safer, more efficient, and environmentally responsible.

8.2.2.7 Stage 6: Handover and Close-Out

The integration of robotics into the Handover and Close-Out stage of the RIBA Plan of Work 2020 presents a paradigm shift in the way construction projects are finalized and delivered. Robotics technology offers several compelling benefits at this critical phase, enhancing efficiency, accuracy, and the overall quality of the handover process. This discussion draws upon existing literature and industry insights to explore these advantages, underpinning the arguments with the principles of Harvard referencing.

Efficiency in Final Inspections and Snagging: Robots equipped with advanced sensors and imaging technologies can significantly expedite the final inspection processes by quickly identifying defects or snags that require attention before handover. These technologies allow for continuous and detailed scanning of the construction site, ensuring that no physical discrepancies or deviations from the project specifications go unnoticed. For instance, drones can perform aerial inspections of hard-to-reach areas, providing high-resolution images that can be analysed to detect issues such as structural anomalies or water leakages.

Documentation and Data Collection: Robotics applications facilitate comprehensive and automated data collection throughout the construction process, culminating in a more streamlined and organized handover process. Robots can autonomously document the as-built conditions of a project, creating detailed records that are essential for the operation and maintenance of the building. This documentation is crucial for facility managers and owners to understand the intricacies of the building systems and for ensuring compliance with regulatory standards.

Enhanced Quality Assurance: The precision and consistency of robotics technology contribute to higher-quality outcomes, a critical aspect during the Handover and Close-Out stage. Robots can perform tasks with a level of accuracy that significantly reduces the risk of human error, ensuring that the finished project adheres to the highest quality standards. This precision is particularly beneficial in tasks that require meticulous attention to detail, such as the installation of mechanical and electrical systems or the application of finishes.

Safety Improvements: Utilizing robots for inspections and final adjustments minimizes the need for human workers to engage in potentially hazardous activities, especially in incomplete or structurally complex areas. This not only reduces the risk of accidents and injuries in the final stages of the project but also ensures compliance with health and safety regulations, safeguarding the wellbeing of the workforce.

Sustainability and Waste Reduction: Robotics can also play a pivotal role in minimizing waste during the Handover and Close-Out phase. By precisely identifying and quantifying leftover materials, robots can facilitate their efficient reuse or recycling, contributing to more sustainable construction practices (Lee et al., 2022). This approach aligns with the growing industry focus on reducing the environmental impact of construction activities.

Accuracy and Quality Control: Accuracy and quality control during the Handover and Close-Out stages are enhanced by robotics. Automated quality checks by robots have been shown to reduce defects streamlining the handover process and ensuring client satisfaction.

In conclusion, the application of robotics in the Handover and Close-Out stage of the RIBA Plan of Work 2020 offers significant benefits, from improving operational efficiencies and quality assurance to enhancing safety and sustainability. As the construction industry continues to evolve, the adoption of robotics technologies represents a forward-looking strategy that can address some of the traditional challenges associated with project handover processes

8.2.2.8 Stage 7: Use

In the context of the RIBA Plan of Work 2020, Stage 7: In Use, the application of robotics offers transformative benefits, enhancing the operational efficiency, sustainability, and adaptability of buildings post-construction. The integration of robotics within this stage reflects an innovative approach to facility management, maintenance, and user interaction, promising to redefine the conventional paradigms of building usage and performance evaluation.

Operational Efficiency and Maintenance: One of the primary benefits of robotics in the 'In Use' stage is the significant enhancement of operational efficiency and maintenance processes. Robots, equipped with advanced sensors and artificial intelligence (AI), can perform routine maintenance tasks, diagnostics, and repairs with greater precision and less downtime compared to traditional methods (Bock and Linner, 2015). For instance, robotic systems can continuously monitor the building's structural health, identifying potential issues before they escalate into costly repairs. This proactive approach to maintenance can extend the lifespan of building components, ensuring optimal performance throughout the building's lifecycle.

Sustainability and Energy Management: Robotics also play a crucial role in advancing sustainability and energy management within buildings. Automated systems can optimize energy consumption by adjusting lighting, heating, ventilation, and air conditioning (HVAC) systems in real-time based on occupancy and environmental conditions. This intelligent management of resources not only reduces the building's carbon footprint but also significantly lowers operational costs, contributing to the building's overall sustainability.

Enhanced User Experience and Safety: Furthermore, robotics enhance the user experience and safety within the building. Robotic assistants can provide guidance and information to occupants, improving accessibility and convenience (Wang et al., 2018). Additionally, in emergency situations, robots equipped with navigation and communication technologies can facilitate efficient evacuation processes, ensuring the safety of the building's occupants.

Adaptability and Space Utilization: The use of robotics in Stage 7 enhances adaptability and efficient space utilization. Robotic systems are capable of rearranging modular components or furniture to accommodate varying uses or events within a building, thereby maximizing versatility and functionality. This adaptability is particularly advantageous in dynamic environments, such as commercial or educational settings, where space requirements frequently shift.

Operational Benefits in the Use Stage: The advantages of robotics extend into the Use stage, where ongoing building maintenance and operations can be significantly optimized. Robots contribute to energy savings and reduced operational costs by improving building efficiency. These enhancements support long-term sustainability and cost-effectiveness, making robotics a valuable asset for efficient building management.

8.2.3 Benefits of Drones in Construction Project Delivery

The benefits associated with the use of drones in construction projects are many and have been extensively examined in many literature (e.g., Albeaino and Gheisari, 2021; Alsamarraie et al., 2022). Some examples of benefits include capability to access unreachable or unsafe areas, and their ability to perform tasks safely and time-efficiently (Gheisari and Esmaeili, 2019; Zhou and Gheisari, 2018). A study by Park et al. (2018) revealed that the global warming potential (GWP) per 1 km delivery by drone was one-sixth that of motorcycle delivery, and the particulates produced by drone delivery were half that of motorcycle delivery. The study further showed that the actual environmental impact reduction in consideration of the delivery distance was 13 times higher in a rural area than in an urban area (Park et al., 2018).

8.2.3.1 Stage 0: Strategic Definition

The application of drone technology during the Strategic Definition stage of the RIBA Plan of Work 2020 offers several compelling benefits. Drones, or unmanned aerial vehicles (UAVs), have transformed various industries, including construction, by providing a novel means of data collection, site analysis, and monitoring. At the Strategic Definition stage (Stage 0), the primary focus is on understanding the client's business case, strategic brief, and assessing the feasibility of the project. The use of drones at this early phase can enhance decision-making processes, provide accurate site assessments, and facilitate the development of a more informed strategic brief.

Enhanced Site Analysis and Selection: Drones offer a unique aerial perspective, enabling comprehensive site surveys and analysis without the limitations of ground-based observation. They can quickly cover large areas, providing high-resolution images and videos that help stakeholders understand the site's topography, access routes, and any potential obstacles. This capability is crucial for accurately assessing the suitability of potential sites, identifying environmental constraints, and planning logistics.

Accurate Data Collection for Feasibility Studies: The accuracy of data collected through drones, equipped with sensors and cameras, supports more precise and detailed feasibility studies. Drones can gather topographic data, monitor environmental conditions, and assess existing infrastructure, which is essential for understanding the project's scope and potential challenges. These data aid in the development of a strategic brief that is grounded in realistic assessments of the project's viability.

Stakeholder Engagement and Communication: Drones can also play a significant role in enhancing stakeholder engagement. By providing visual representations of the site and its context, drones help stakeholders visualize the project in its early stages, fostering a clearer understanding and facilitating informed discussions. This can lead to more effective communication and alignment of the project objectives with the strategic goals of the client.

Environmental and Contextual Analysis: Environmental sustainability is a critical consideration in contemporary construction projects. Drones contribute to a thorough environmental analysis by mapping out vegetation, water bodies, and wildlife habitats, thus informing strategies for minimizing environmental impact. Moreover, by analysing the site's context, including neighbouring buildings and land use, drones help in developing a project that harmonizes with its surroundings.

Risk Assessment and Mitigation Planning: Identifying and mitigating risks at the earliest stage of a project can save time, resources, and potential legal complications. Drones offer a safe and efficient means of identifying site-specific risks, such as unstable ground, flooding areas, or difficult access points. This information is invaluable for planning mitigation strategies and ensuring the project's feasibility from a risk management perspective.

In conclusion, the integration of drone technology into the Strategic Definition stage of the RIBA Plan of Work 2020 enhances the quality and efficiency of site analysis, feasibility studies, and strategic planning. By providing detailed, accurate data and visual insights, drones support the development of a robust strategic brief that aligns with the client's objectives and the project's environmental, logistical, and regulatory requirements. As such, drones are not merely a technological tool but a strategic asset in the early planning phases of construction projects.

Drones facilitate early site analysis by capturing comprehensive data on topography, existing structures, and vegetation, aiding in feasibility studies and initial project considerations. This capability provides a cost-effective method for collecting extensive site data, informing strategic decisions and planning (Albeaino and Gheisari, 2021).

8.2.3.2 Stage 1: Preparation and Brief

The integration of drone technology during Stage 1: Preparation and Brief of the RIBA Plan of Work 2020 offers a multitude of benefits that significantly enhance the preliminary phase of construction projects. Drones, with their advanced imaging and data collection capabilities, provide a novel approach to site analysis, topographical surveying, and environmental assessments, thereby facilitating a more informed and efficient planning process.

Enhanced Site Analysis and Topographical Surveying: Drones equipped with high-resolution cameras and LiDAR technology enable detailed aerial surveys of the project site. This capability allows for the rapid assessment of site conditions, including topography, existing structures, and natural features, with a degree of accuracy and detail that was previously unattainable with traditional surveying methods (Vishweshwaran and Sujatha, 2024). The aerial perspective offered by drones provides comprehensive insights into the site, aiding architects and planners in understanding the spatial characteristics and constraints early in the project lifecycle.

Accurate Environmental Assessments: The environmental assessment of a project site is crucial in the early stages of planning to identify potential ecological impacts and ensure compliance with regulatory requirements. Drones facilitate detailed environmental surveys by capturing data on vegetation, wildlife habitats, and water bodies, enabling a more accurate evaluation of the environmental context (Sharma and Naik, 2024). This information is vital for developing a sustainable project brief that aligns with environmental conservation goals.

Improved Communication and Stakeholder Engagement: Visual data collected by drones can be transformed into detailed 3D models and visualizations, improving communication among project stakeholders (Salem et al., 2024). By providing vivid representations of the site and its context, drones help stakeholders better understand the project's scope and constraints, fostering more effective collaboration and decision-making during the preparation and brief stage.

Cost and Time Efficiency: Conducting site surveys and environmental assessments with drones is significantly faster and often less expensive than traditional methods. Drones can cover large areas in a short period, reducing the time and labour required for site analysis (Choi et al., 2023). This efficiency not only accelerates the project timeline but also reduces preliminary costs, contributing to a more streamlined and cost-effective project initiation phase.

Utilizing drones for detailed aerial surveys supports the preparation phase, enhancing understanding of the site's conditions. This is instrumental in developing the project brief, ensuring alignment with the physical context and constraints identified (Alsamarraie et al., 2022).

8.2.3.3 Stage 2: Concept Design

Drones, or UAVs, have emerged as a transformative technology in the construction industry, particularly in the early stages of project development such as Stage 2: Concept Design of the

RIBA Plan of Work 2020. Their application during this stage offers several benefits, enhancing the design process through improved data collection, visualization capabilities, and stakeholder engagement. This discussion will explore these benefits, underpinned by the principles of Harvard referencing.

Enhanced Site Analysis and Data Collection: Drones provide architects and design teams with the capability to conduct comprehensive site surveys quickly and accurately. They can capture high-resolution aerial imagery and topographical data, facilitating a detailed analysis of the site's features, boundaries, and potential constraints. This information is crucial for informed decision-making during the concept design stage. As noted by Bayomi and Fernandez (2023), drones streamline the site analysis process, enabling designers to assess environmental factors, such as sunlight and wind patterns, which significantly influence the conceptual design.

Improved Visualization and Design Innovation: The use of drones extends beyond mere data collection; they also offer advanced visualization capabilities. Through aerial photography and 3D mapping, drones help create detailed and immersive representations of the site, allowing design teams to explore different design scenarios and spatial relationships more effectively. Farouk et al. (2024) highlights how these visualizations can be integrated into virtual reality (VR) environments, providing a powerful tool for design innovation and experimentation at the conceptual stage.

Enhanced Stakeholder Engagement and Communication: Drones facilitate a more engaging and informative communication process with stakeholders (Salem et al., 2024). By providing real-time aerial footage and 3D models of the proposed design in the context of the actual site, stakeholders can gain a clearer understanding of the project scope and design intentions. This aids in building consensus and addressing concerns early in the design process.

Risk Reduction and Preliminary Planning: The application of drones during the concept design stage also plays a crucial role in identifying and mitigating potential risks. Aerial surveys can uncover site-specific challenges, such as difficult terrain, existing structures, or environmental sensitivities, that may impact the design and construction process. By identifying these issues early, design teams can develop strategies to mitigate risks, leading to more accurate planning and budgeting. This proactive approach to risk management can lead to significant cost savings and a smoother project execution phase.

During concept design, drones offer detailed 3D mappings and aerial views, aiding in the visualization of the project within its environmental context. These insights facilitate addressing design challenges early, leading to more accurate and contextually integrated architectural concepts (Gheisari and Esmaeili, 2019).

8.2.3.4 Stage 3: Spatial Coordination

During Stage 3: Spatial Coordination of the RIBA Plan of Work 2020, the integration of drones presents a transformative potential for the construction industry. This stage, crucial for the development of design through detailed spatial coordination, benefits significantly from the application of drone technology. Drones offer a unique vantage point and capabilities that can enhance the efficiency, accuracy, and comprehensiveness of spatial coordination efforts.

Efficiency in Data Collection: Drones facilitate rapid and efficient collection of aerial data, enabling comprehensive site analysis and topographical mapping with much greater speed than traditional methods (Barnes and Davies, 2014). This capability is especially beneficial during the spatial coordination stage, where understanding the physical context is crucial for the integration of design elements. Drones can quickly survey a construction site, providing real-time data that helps in making informed decisions without the delays associated with ground-based survey methods.

Enhanced Accuracy and Detail: The high-resolution imagery and 3D mapping capabilities of drones provide detailed insights into the site conditions, offering a level of detail that supports the precise coordination of spatial elements. This accuracy is vital for identifying potential issues early in the design process, allowing for adjustments before construction begins, thereby reducing costly errors and rework.

Improved Collaboration and Communication: Drones contribute to enhanced collaboration among project stakeholders by providing visual data that can be easily shared and discussed. During spatial coordination, the ability to visualize the site and design elements in their actual context helps in aligning the expectations and understanding among architects, engineers, and clients. This improved communication supports a more integrated and cohesive design process.

Risk Mitigation: By enabling thorough site analysis and monitoring, drones help identify potential risks and obstacles early in the design phase. This proactive approach to risk management is crucial during spatial coordination, as it allows for the design to adapt to site-specific challenges, thereby mitigating risks before they impact the construction phase.

Sustainability and Environmental Considerations: Drones also play a role in assessing the environmental impact of a project, aiding in the sustainable integration of designs with their natural surroundings. This assessment can lead to design adjustments that minimize environmental impact and enhance the sustainability of the construction project.

To conclude, the application of drones during the Spatial Coordination stage of the RIBA Plan of Work 2020 offers significant benefits in terms of efficiency, accuracy, collaboration, risk mitigation, and sustainability. These advantages underscore the value of integrating drone technology into the architectural and construction processes, providing a foundation for more informed, efficient, and effective project development.

8.2.3.5 Stage 4: Technical Design

The integration of drone technology into Stage 4: Technical Design of the RIBA Plan of Work 2020 offers significant benefits, enhancing the precision, efficiency, and comprehensiveness of technical design processes. Drones, with their advanced imaging and data collection capabilities, can provide architects, engineers, and construction professionals with detailed insights into the project site and its surroundings, facilitating a more informed and accurate design phase. This section discusses the specific advantages of drone applications in the Technical Design stage, supported by relevant references.

Enhanced Site Analysis: Drones can capture high-resolution aerial imagery and topographical data, enabling detailed site analysis before finalizing technical designs. This capability allows for the identification of potential issues, such as topographical challenges, existing infrastructure, and environmental constraints, which can significantly impact the design. By providing a comprehensive overview of the site, drones help in creating more accurate and sustainable designs.

Improved Accuracy and Detail in Design Documents: The precision of drone-collected data translates into improved accuracy in design documents and drawings. High-resolution images and 3D models generated from drone data can be integrated into BIM systems, enhancing the quality and detail of technical design documents. This integration not only facilitates better communication among project stakeholders but also reduces the risk of errors and discrepancies in design documents.

Efficient Monitoring and Updating of Design Changes: Drones offer a dynamic tool for monitoring ongoing site conditions and implementing design changes in real-time. As the technical design phase often requires adjustments based on site conditions and other factors, drones can quickly capture updated site information, allowing design teams to make informed decisions and adjustments without significant delays. This efficiency in monitoring and updating design changes can lead to cost savings and ensure the project remains on schedule.

Enhanced Collaboration and Stakeholder Engagement: The visual and data-rich output from drone operations can improve collaboration among project stakeholders. By providing clear, detailed aerial views and 3D models of the project site, drones help stakeholders visualize the technical design in the context of the actual site conditions. This enhanced visualization can facilitate more effective communication and decision-making among architects, engineers, clients, and contractors, leading to a more cohesive and aligned project team.

Risk Mitigation: Drones contribute to risk mitigation during the technical design stage by identifying potential issues early in the design process. The detailed site data collected by drones can reveal risks related to site access, construction feasibility, and environmental impact, allowing the design team to address these issues proactively. By mitigating risks early, drones help prevent costly design modifications and project delays.

Drones play a crucial role in verifying site conditions against the technical designs, ensuring project plans align with actual site conditions. This process helps in identifying discrepancies early, preventing costly adjustments during construction (Alsamarraie et al., 2022).

8.2.3.6 Stage 5: Manufacturing and Construction

The application of drone technology within Stage 5: Manufacturing and Construction of the RIBA Plan of Work 2020 offers numerous benefits, revolutionizing traditional construction practices through enhanced efficiency, safety, and data collection capabilities. Drones, also known as UAVs, have emerged as pivotal tools in modern construction processes, providing a unique vantage point that was previously difficult or expensive to obtain.

Enhanced Project Monitoring and Management: Drones facilitate real-time monitoring and management of construction sites, enabling project managers to obtain aerial views that offer comprehensive insights into the progress of construction activities. This capability allows for the rapid identification of discrepancies between the planned and actual progress, leading to more informed decision-making and timely adjustments to project execution. By offering a bird's-eye view, drones enable project managers to oversee large construction sites efficiently, ensuring that projects adhere to their timelines and budgets.

Improved Safety: Safety on construction sites is significantly enhanced through the use of drones. By conducting aerial surveys, drones can identify potential hazards and monitor compliance with safety protocols without putting human lives at risk (Bolaji et al., 2024). This application is particularly valuable in accessing hard-to-reach areas, such as high-rise structures or hazardous zones, where traditional methods of inspection might expose workers to risk. Drones thus play a crucial role in pre-emptively addressing safety concerns and reducing the likelihood of accidents on site.

Accurate Data Collection and Surveying: Drones contribute to the accuracy and efficiency of data collection and surveying processes. Equipped with high-resolution cameras and sensors, drones can capture detailed images and generate precise topographical maps and 3D models of construction sites (Choi et al., 2024). These data are invaluable for assessing site conditions, planning construction activities, and conducting volume calculations for materials. The ability to rapidly collect and process data ensures that construction projects can proceed with a clear understanding of the site conditions, reducing the risk of errors and rework.

Enhanced Communication and Stakeholder Engagement: The visual documentation captured by drones facilitates improved communication among project stakeholders. Aerial imagery and video footage provide a tangible means of demonstrating construction progress to clients, investors, and other interested parties, fostering transparency and trust. This visual evidence can also support marketing efforts, showcasing the scale and complexity of construction projects to potential clients and the broader public.

The integration of drone technology in the Manufacturing and Construction stage of the RIBA Plan of Work 2020 offers transformative benefits, driving efficiency, safety, and accuracy in construction practices. As the construction industry continues to evolve, the adoption of drones is set to play an increasingly significant role in shaping the future of construction project management and execution.

The construction phase sees drones monitoring progress, ensuring safety compliance, and identifying potential issues promptly. Their ability to safely access difficult areas maintains project timelines and budgets (Albeaino and Gheisari, 2021).

8.2.3.7 Stage 6: Handover and Close-Out

The integration of drone technology during the Stage 6: Handover and Close-Out of the RIBA Plan of Work 2020 offers a myriad of benefits that streamline and enhance the efficiency, accuracy, and comprehensiveness of the construction handover process. This stage, pivotal for ensuring that the project meets its outlined specifications and quality before the final handover to the client, can significantly benefit from the capabilities provided by drones. The following discussion outlines these advantages, supported by relevant literature and industry practices.

Efficiency in Completion and Defect Inspection: Drones provide a rapid and efficient means of conducting final inspections and identifying any remaining defects or work that needs completion. Their ability to quickly cover large areas and access hard-to-reach places allows for comprehensive inspections that might be time-consuming and hazardous if done manually. For instance, aerial surveys can capture detailed images of roofing, façades, and other elevated structures, facilitating a quicker assessment by the project team.

Enhanced Documentation for Handover: Drones contribute to creating detailed and accurate documentation for project handover, including high-resolution images and videos of the completed project. This visual documentation can be invaluable for clients, providing a clear and thorough record of the construction's final state. Such detailed aerial perspectives can also support the preparation of as-built documentation, ensuring that it accurately reflects the completed project.

Improved Communication with Stakeholders: The visual data captured by drones can be shared with clients and stakeholders, offering a transparent and immersive view of the project outcome. This supports clear and effective communication, helping to address any concerns or queries the client may have regarding the project's completion. By providing an aerial overview, stakeholders can better understand the completed work, facilitating a smoother handover process.

Safety and Risk Reduction: Drones minimize the need for manual inspections in potentially hazardous areas, thereby reducing safety risks associated with the handover and close-out process. The ability to conduct aerial inspections remotely means that issues can be identified and resolved without exposing personnel to risks, such as working at heights or in confined spaces.

Cost and Time Savings: By streamlining the inspection process and reducing the need for physical access to difficult areas, drones can lead to significant cost and time savings during the handover stage. The efficiency of drones in capturing comprehensive data reduces the man-hours required for inspections and accelerates the close-out process, allowing for quicker project delivery and reduced costs.

Inspection and Documentation: Drones assist in the final inspection and documentation, offering comprehensive aerial imagery for promotional materials or handover documents. This ensures thorough documentation of all project aspects, facilitating a smooth transition to the building's users (Gheisari and Esmaeili, 2019).

To conclude, the application of drone technology during the Stage 6: Handover and Close-Out of the RIBA Plan of Work 2020 offers compelling benefits that enhance the efficiency, accuracy, and safety of the construction handover process. The ability of drones to provide detailed aerial surveys, coupled with their efficiency and safety advantages, positions them as a valuable tool in modern construction management practices.

8.2.3.8 Stage 7: Use

Drones, or UAVs, have revolutionized many aspects of the construction industry, offering significant benefits from the project's inception through to its operational phase. In the context of Stage 7: In Use, of the RIBA Plan of Work 2020, the application of drones extends beyond construction, contributing to the efficient operation, maintenance, and lifecycle management of buildings. This discussion explores the multifaceted benefits of drone technology in the operational phase of buildings, referencing relevant literature and studies to underline their impact.

Facilitating Regular Inspections and Maintenance: Drones provide a cost-effective and efficient means for conducting regular inspections of buildings, especially in hard-to-reach areas. By enabling aerial views and capturing high-resolution images, drones can help identify maintenance needs, such as roof damage, gutter blockages, or structural issues, without the need for scaffolding or manual access. This not only reduces the risk of accidents but also decreases the time and labour costs associated with traditional inspection methods.

Monitoring Building Performance: Drones equipped with thermal imaging cameras can assess a building's thermal performance, identifying heat losses and inefficiencies (Smith, 2014). This application is crucial for optimizing energy consumption and enhancing the sustainability of buildings in their operational phase. By detecting thermal anomalies, stakeholders can make informed decisions regarding necessary retrofits or adjustments to improve energy efficiency.

Enhancing Security and Surveillance: The use of drones for security purposes in the operational phase of buildings adds a dynamic surveillance capability, offering real-time monitoring of the premises. This can deter criminal activity, ensure the safety of occupants, and provide valuable documentation in case of security breaches. The mobility and flexibility of drones make them an invaluable tool for comprehensive security management.

Supporting Facility Management: Drones can play a significant role in supporting facility management operations by providing aerial overviews of the site, assisting in asset management, and monitoring construction or renovation works within the building premises. This broad overview can help facility managers plan maintenance work more efficiently, manage assets effectively, and ensure smooth operations throughout the building's lifecycle.

Emergency Response and Disaster Recovery: In the event of natural disasters or emergencies, drones can be rapidly deployed to assess damage, navigate through debris, and assist in the recovery process. They provide critical information that can guide emergency response efforts, evaluate the structural integrity of buildings, and prioritize actions to mitigate further risks to occupants and property.

Monitoring: During the building's operational phase, drones offer a sustainable method for monitoring its condition and performance. The reduced environmental impact of using drones is particularly beneficial for ongoing maintenance and management of rural buildings (Park et al., 2018).

In conclusion, the integration of drones in the operational phase of buildings, as outlined in Stage 7 of the RIBA Plan of Work 2020, offers substantial benefits across maintenance, security, energy efficiency, and emergency response domains. Their application enhances the operational efficiency, sustainability, and safety of buildings, underlining the value of drone technology in the post-construction lifecycle of built environments. As the technology continues to evolve, the potential for drones to contribute to the operational and maintenance aspects of buildings will likely expand, further solidifying their role in the future of building management and facility operations.

8.2.4 Benefits of AR/VR/MR

8.2.4.1 Stage 0: Strategic Definition

MR for Initial Project Visualization: At the outset, mixed reality technology offers stakeholders an interactive means to visualize and assess the feasibility of potential projects. By enabling an immersive exploration of virtual models, MR can aid in decision-making and strategic planning, potentially reducing preliminary planning time by up to 15% and increasing stakeholder engagement by 20% (Sacks et al., 2013).

Enhanced Visualization: AR, VR, and MR enable stakeholders to visualize potential project outcomes more effectively, helping to make informed decisions about whether a construction project is the best way to achieve client requirements.

Risk Identification: These technologies allow for the simulation of various scenarios and their impact, aiding in identifying potential project risks early in the strategic phase.

Cost Estimation and Feasibility: VR and MR can simulate different project options, allowing for more accurate cost estimation and feasibility analysis before committing to a particular course of action.

Stakeholder Engagement: AR and VR can provide immersive experiences that help stakeholders, including non-technical clients, to engage with and understand the strategic choices available, leading to better decision-making.

Site Analysis: MR and AR can overlay virtual elements onto real-world environments, assisting in site appraisals and planning evaluations to better understand site constraints and opportunities.

Improved Communication: These technologies facilitate clearer communication among the project team and with clients, ensuring everyone has a consistent understanding of the strategic objectives.

Efficient Option Evaluation: AR and VR enable rapid prototyping of different strategic options, making it easier to compare and contrast them without the need for physical mock-ups or extensive preliminary designs.

Improved Security: MR walkthroughs can be used early on to identify potential safety concerns, which impacts the overall project strategy by reducing risks before construction begins.

8.2.4.2 Stage 1: Preparation and Brief

Enhanced Project Planning with MR: During the preparation and brief stage, MR plays a pivotal role in refining project objectives and requirements. Utilizing MR for detailed project exploration and planning can help identify up to 75% of design conflicts early, leading to a reduction in overall project costs and schedules by approximately 10%, by ensuring that all stakeholders have a clear and shared understanding of the project goals (Kim et al., 2017).

Enhanced Briefing Process: AR, VR, and MR can create immersive environments that help clients better understand and refine their project briefs, ensuring that their needs and expectations are clearly communicated and understood by the design team.

Improved Stakeholder Communication: These technologies enable more effective communication with stakeholders by visualizing project concepts in 3D, making it easier to gather feedback and make adjustments early in the process.

Accurate Spatial Planning: AR and MR can overlay virtual layouts onto the actual site, allowing for real-time spatial planning and helping to assess the feasibility of the proposed design against site constraints.

Cost and Time Efficiency: By visualizing potential designs in VR or MR, the project team can identify and resolve potential issues early, reducing the need for costly revisions later in the project.

Interactive Design Prototyping: VR and MR enable interactive design prototyping, allowing clients and stakeholders to virtually walk through proposed designs and suggest changes before the design process moves forward.

Better Decision-Making: These technologies provide a clear, detailed understanding of the project's scope and requirements, leading to more informed decision-making by both clients and the project team.

Alignment of Expectations: AR and VR ensure that all parties have a consistent and accurate understanding of the project's goals, reducing the risk of misaligned expectations and ensuring that the brief accurately reflects the client's vision.

Using AR, VR, and MR during Stage 1 helps in creating a more precise and shared understanding of the project requirements, leading to a smoother design process in subsequent stages.

Better project planning (MR): MR allows for detailed exploration of the entire project during the preparation and briefing stage, ensuring all project components are well understood and planned.

8.2.4.3 Stage 2: Concept Design

Client Engagement Through VR: VR is invaluable in enhancing client understanding and engagement during the concept design phase. The immersive experience provided by VR helps clients visualize complex architectural concepts, increasing client satisfaction by up to 20% and decision-making efficiency by 25% through improved feedback and preferences communication (Ahmed, 2019).

Enhanced Design Visualization: VR and MR provide immersive, 3D visualizations of the concept design, allowing clients and stakeholders to experience the proposed design in a lifelike environment, facilitating better understanding and engagement.

Rapid Design Iteration: AR, VR, and MR enable quick changes to the design, allowing designers to test and present multiple iterations efficiently, helping to refine the concept design based on real-time feedback.

Improved Spatial Coordination: MR can overlay the concept design onto the physical site, helping to assess how the proposed design fits within the existing environment and making it easier to identify spatial issues early on.

Enhanced Client and Stakeholder Communication: These technologies make it easier to communicate complex design ideas, ensuring that all parties have a clear and shared understanding of the concept, which helps in obtaining faster approvals and aligning expectations.

Realistic Simulation of Environmental Conditions: VR can simulate different environmental conditions (e.g., lighting, weather, acoustics) within the concept design, allowing for more informed decisions about the design's impact on and interaction with its surroundings.

Increased Design Confidence: By experiencing the design in VR or MR, clients can make more confident decisions, reducing the likelihood of significant changes or dissatisfaction later in the project.

Integration of User Experience: VR allows stakeholders to explore the design from a user's perspective, ensuring that the concept design meets the functional and experiential needs of future users.

Collaboration Across Disciplines: AR, VR, and MR facilitate better collaboration among architects, engineers, and other consultants by providing a shared platform to view and discuss the evolving design, leading to more integrated and cohesive outcomes.

Enhancing Client's Understanding: Using VR to help clients visualize and immerse themselves in the project during the concept design phase to improve engagement and understanding.

Detection of Errors: Initial walkthroughs in VR can be carried out during concept design to identify potential problems or design flaws before moving further.

8.2.4.4 Stage 3: Developed Design

VR for Error Detection: Utilizing VR for virtual walkthroughs at the developed design stage enables early detection of design errors. This proactive approach can lead to a 30% reduction in design errors and decrease the need for later stage changes by up to 40%, significantly saving time and costs associated with reworks (Kim et al., 2017).

Detailed Design Validation: VR and MR allow for immersive walkthroughs of the developed design, enabling the project team and clients to validate detailed aspects of the design, ensuring that all elements meet the project requirements and function as intended.

Enhanced Coordination Across Disciplines: AR and MR facilitate the overlay of different design disciplines (e.g., architecture, structure, MEP) in a shared virtual space, ensuring that all systems are well-coordinated and reducing the risk of clashes or integration issues.

Efficient Design Refinement: These technologies enable quick and precise adjustments to the developed design based on real-time feedback, allowing for efficient refinement before finalizing the design.

Improved Stakeholder Engagement: VR provides stakeholders with an immersive experience of the developed design, helping them to better understand and engage with the project, leading to more informed and aligned decision-making.

Simulation of User Experience: VR can simulate the end-user experience within the developed design, ensuring that the design meets the intended functional and aesthetic objectives, and providing valuable insights into how the space will be used.

Real-time Environment Interaction: MR allows the developed design to be visualized within the actual site context, helping to assess how the design interacts with the surrounding environment and making necessary adjustments to optimize the design's fit and impact.

Enhanced Communication of Design Intent: AR and VR help in clearly communicating the developed design intent to construction teams and other stakeholders, ensuring that everyone has a consistent understanding of the design details.

Early Detection of Potential Issues: These technologies allow for the early identification of potential design issues, such as spatial conflicts or user experience problems, reducing the likelihood of costly changes during later stages of the project.

Schedule and Cost Optimization: VR models can be explored in this phase to optimize scheduling and cost by planning work efficiently and avoiding costly delays and errors.

Avoidance of Reworks: During developed design, VR helps identify errors early, ensuring that changes can be made before the construction begins, reducing the likelihood of reworks later.

Utilizing AR, VR, and MR in Stage 3 ensures that the developed design is thoroughly vetted, well-coordinated, and ready for the technical detailing required in subsequent stages, leading to a smoother design process and better project outcomes

8.2.4.5 Stage 4: Technical Design

Improving Collaboration with VR: VR technology fosters enhanced collaboration among project teams by allowing them to review and interact with the design in a shared virtual space. This collaborative environment can improve project communication efficiency by 35% and reduce misunderstandings and errors by 25%, streamlining the technical design process (Vivian, 2022).

Precise Design Detailing: VR and MR enable designers to explore and refine technical details in an immersive 3D environment, ensuring that all components are accurately developed and integrated.

Improved Design Coordination: AR and MR allow for real-time visualization of how different technical elements, such as structural, mechanical, and electrical systems, interact within the design, reducing the risk of conflicts and ensuring seamless integration.

Enhanced Collaboration: These technologies provide a shared virtual space for architects, engineers, and contractors to collaborate on technical details, facilitating clear communication and reducing misunderstandings.

Virtual Prototyping: VR enables virtual prototyping of complex elements, allowing the project team to test and refine the technical design before physical construction begins, reducing the likelihood of errors and rework.

Efficient Review and Approval Process: AR and VR streamline the review process by allowing stakeholders to experience and assess the technical design in a detailed, interactive format, leading to quicker approvals and fewer revisions.

Simulation of Construction Methods: MR can simulate construction sequences and methods, helping to identify potential challenges and optimize the construction process, ensuring that the design is practical and buildable.

Real-time Issue Resolution: These technologies allow for the immediate identification and resolution of technical issues as they arise, reducing delays and ensuring that the design is fully resolved before construction begins.

Accurate Documentation: AR and VR assist in generating accurate, detailed documentation by visualizing how components fit together in the built environment, ensuring that all technical drawings and specifications are correct and comprehensive.

Enhanced Stakeholder Confidence: By providing a clear, immersive view of the technical design, these technologies increase stakeholder confidence in the design's feasibility and quality, reducing the risk of last-minute changes or concerns.

Improve Collaboration: VR allows all stakeholders to engage with the project from various perspectives, improving communication and coordination. It also helps in familiarizing on-site workers with the project.

Resource Management (MR): MR walkthroughs can be used to assess resources and potential problems before finalizing the technical design, ensuring the project proceeds without issues.

In Stage 4, AR, VR, and MR play a crucial role in refining and finalizing the technical design, ensuring that all details are accurately developed, well-coordinated, and ready for the construction phase.

8.2.4.6 Stage 5: Manufacturing and Construction

VR for Schedule and Cost Optimization: During construction, VR's ability to foresee and prevent construction errors can decrease project delays by up to 20% and reduce unforeseen costs by 15%. Additionally, VR-based training modules can enhance on-site worker efficiency by 10%, further optimizing the construction phase (Ahmed, 2019; Sacks et al., 2013).

Resource Management Through MR: MR technology provides actionable insights into resource allocation and workflow optimization during construction, improving project delivery efficiency by 10–15% and ensuring adherence to schedules and budgets (Kim et al., 2017).

Improved Construction Accuracy: AR can overlay digital models onto the physical construction site, guiding workers with precise placement and installation of components, reducing errors, and ensuring that the construction aligns exactly with the design.

Enhanced Quality Control: VR and MR allow for virtual inspections of construction progress, enabling early detection of discrepancies or defects, which can be corrected before they become larger issues.

Streamlined Communication: These technologies facilitate clear, real-time communication between on-site workers, off-site teams, and stakeholders, ensuring that everyone is aligned and any issues are quickly addressed.

Efficient Training and Simulation: VR provides immersive training environments where workers can practice and refine construction techniques before applying them on-site, leading to increased productivity and fewer mistakes.

Optimized Construction Sequencing: MR can simulate and visualize construction sequences, helping project managers plan and optimize the order of operations to improve efficiency and reduce delays.

Real-time Monitoring and Feedback: AR enables real-time monitoring of construction activities, providing immediate feedback to workers and supervisors, helping to maintain high standards and adherence to the design.

Safer Construction Environment: AR and MR can highlight potential hazards on-site, improving safety by guiding workers around risks and ensuring compliance with safety protocols.

Efficient Assembly and Installation: MR assists in the assembly of prefabricated components by providing visual guides that ensure each piece is correctly positioned, speeding up the construction process and reducing the need for rework.

Stakeholder Engagement: VR allows stakeholders to virtually tour the construction site at different stages, providing them with a clear understanding of progress and any necessary adjustments, which enhances transparency and trust.

Accurate Progress Tracking: AR and VR tools can be used to compare the current construction state with the planned design, ensuring that the project stays on schedule and identifying any deviations that need to be addressed.

Avoidance of Reworks: VR continues to be useful in the construction phase to prevent reworks by ensuring that the construction proceeds according to the virtual model.

Improve Collaboration (Training On-Site Workers): VR training helps on-site workers familiarize themselves with the environment, improving teamwork and communication during the construction process.

In Stage 5, AR, VR, and MR enhance the construction process by improving accuracy, efficiency, safety, and communication, leading to higher-quality outcomes and smoother project execution.

8.2.4.7 Stage 6: Handover and Close-Out

MR for Improved Security: Utilizing MR for security walkthroughs during handover can identify potential safety issues, reducing post-construction safety incidents by 50% and ensuring a safer environment for end-users while minimizing liability risks (Kim et al., 2017).

Comprehensive Final Inspections: VR and MR allow for detailed virtual walkthroughs of the completed project, enabling thorough final inspections to ensure that all aspects of the design have been accurately implemented and meet the required standards.

Enhanced Client Demonstrations: AR and VR provide immersive experiences for clients during handover, allowing them to explore and understand the building's features, systems, and operations in a more interactive and engaging manner.

Efficient Documentation and Training: AR can be used to overlay instructions and operational guides directly onto the physical space, making it easier for clients and facility managers to understand building systems and perform necessary maintenance tasks.

Improved Defect Identification and Rectification: MR can help identify defects or areas requiring attention by overlaying the as-built model with the design model, ensuring that any discrepancies are addressed before final handover.

Streamlined Handover Process: These technologies facilitate a more efficient handover process by allowing for real-time updates and modifications to digital models, ensuring that all handover documentation is accurate and up-to-date.

Enhanced Client Satisfaction: By providing an immersive and interactive handover experience, VR and AR can increase client satisfaction by ensuring they fully understand and appreciate the final product, reducing the likelihood of post-handover issues.

Detailed Asset Information Management: MR can link physical building components to their digital counterparts, providing detailed, easily accessible information about each asset, which is invaluable for future maintenance and management.

Facilitated Post-Occupancy Evaluation: VR can simulate different occupancy scenarios, helping to evaluate building performance in use and gather valuable feedback for future projects or improvements.

Efficient Defects Liability Period Management: AR can track and document any issues that arise during defect's liability period, ensuring they are resolved promptly and reducing the administrative burden on both the client and the project team.

Smooth Transition to Building Use: AR and MR assist in training building users on systems and operations, ensuring a smooth transition to full operational use without disruptions or misunderstandings.

Improved Security: The early identification of safety issues via MR walkthroughs ensures a safer handover process as construction hazards are minimized.

In Stage 6, AR, VR, and MR significantly enhance the handover and close-out process by improving inspection accuracy, client engagement, documentation, and overall satisfaction, ensuring a smooth and successful project completion.

8.2.4.8 Stage 7: Use

Ongoing Communication and Visualization with AR and VR: AR and VR technologies facilitate maintenance, renovations, and operational adjustments post-construction by providing detailed visualizations of building components. Their application can improve maintenance efficiency by 20% and reduce operational costs by up to 10%, enhancing decision-making and building lifecycle management (Ahmed, 2019).

Enhanced Building Maintenance: AR provides real-time, on-site visual guides and instructions for maintenance tasks, improving efficiency and accuracy in maintaining building systems and reducing downtime.

Optimized Facility Management: MR integrates digital asset information with the physical environment, allowing facility managers to access detailed data on building components, streamlining operations, and making informed decisions.

Interactive User Training: VR offers immersive training environments for building users and maintenance staff, ensuring they are well-equipped to operate and manage the building effectively.

Real-Time Performance Monitoring: AR and MR enable continuous monitoring of building systems, providing real-time data on performance metrics, energy use, and occupant comfort, leading to better building management and reduced operational costs.

Improved Space Utilization: VR can simulate different usage scenarios within the building, helping users optimize space utilization and adapt the building layout to evolving needs.

User Experience Enhancement: AR can enhance user experience by providing interactive information about the building's features, history, and services, enriching the environment for occupants and visitors.

Proactive Issue Detection: MR facilitates the early detection of potential issues within building systems by overlaying real-time data onto physical assets, allowing for proactive maintenance and reducing the likelihood of system failures.

Efficient Energy Management: VR and MR can simulate energy usage patterns and suggest optimizations, helping to achieve sustainability goals and reduce the building's carbon footprint.

Streamlined Post-Occupancy Evaluations: VR allows for immersive post-occupancy evaluations, enabling stakeholders to virtually revisit the building and assess its performance and user satisfaction over time.

Long-term Asset Management: MR supports the long-term management of the building by integrating digital twins with physical assets, providing a comprehensive view of the building's lifecycle and facilitating informed decisions on renovations, upgrades, or repurposing.

Enhancing Client's Understanding: Even post-construction, VR can be used to engage clients, ensuring they understand the final layout and operational aspects of the building.

In Stage 7, AR, VR, and MR significantly improve the ongoing use, management, and optimization of the building, enhancing user experience, operational efficiency, and long-term sustainability.

8.2.5 *Benefits of 3D Printers*

Some of the benefits of 3D printing in construction will be discussed in the ensuing paragraphs.

8.2.5.1 *Stage 0: Strategic Definition*

At this foundational stage, project stakeholders establish the business case and strategic approach, focusing on understanding the client's needs, project feasibility, and overarching goals.

Environmental Benefit: Leveraging the environmental advantages of 3D printing from the outset can prioritize sustainability. The ability to reduce material waste and carbon footprint through 3D printing supports strategic decisions aligning project goals with sustainability targets, setting a precedent for environmentally conscious construction methods.

Rapid Conceptualization: 3D printers allow for the quick creation of physical models, helping clients and stakeholders visualize potential project options early in the strategic phase, aiding in decision-making.

Enhanced Stakeholder Engagement: Physical models created through 3D printing can be used to engage stakeholders more effectively, making it easier to communicate abstract concepts and gather feedback.

Cost and Feasibility Analysis: 3D-printed models provide tangible representations of different strategic options, enabling more accurate assessments of cost, feasibility, and potential project risks before committing to a specific direction.

Early Detection of Design Challenges: Creating physical models in Stage 0 can help identify potential design challenges or spatial constraints early, allowing for adjustments before moving forward with detailed planning.

Tangible Comparative Analysis: 3D printing enables the production of multiple design options, allowing for a direct, side-by-side comparison of different strategies, which facilitates a more informed decision-making process.

Increased Client Confidence: Providing clients with physical models of potential solutions increases their confidence in the strategic decisions being made, ensuring alignment before moving to the next stage.

Facilitated Communication: 3D-printed models serve as effective tools for communication among project teams, helping to clarify the strategic vision and ensuring that all parties have a clear understanding of the project's direction.

Environmental Benefit: The overall reduction in material wastage and the environmental impact of using 3D printing, such as improved fuel efficiency and reduced $CO2$ emissions, can influence the strategic decision to incorporate 3D printing in construction projects.

Using 3D printers in Stage 0 helps in making the strategic definition process more tangible, precise, and collaborative, leading to better-informed decisions and a clearer path forward for the project.

8.2.5.2 *Stage 1: Preparation and Brief*

Here, the project brief is developed, detailing the client's requirements and objectives, including functionality, aesthetics, and sustainability.

Flexible Design: The design flexibility offered by 3D printing influences the project brief by allowing for innovative solutions to complex architectural features. Rapid prototyping capabilities enable stakeholders to visualize and refine project objectives, encouraging creative exploration of forms and structures.

Enhanced Brief Development: 3D-printed models can bring the project brief to life, allowing clients and stakeholders to see and feel the physical form of initial design concepts, which helps in refining and finalizing the brief.

Improved Stakeholder Communication: Physical models created by 3D printers facilitate better communication with stakeholders, making it easier to convey complex ideas and ensuring that everyone has a shared understanding of the project's scope and objectives.

Informed Decision-Making: 3D-printed prototypes allow clients to visualize spatial relationships and scale, enabling more informed decisions regarding layout, functionality, and design features before moving into detailed design stages.

Cost Estimation and Feasibility: By producing physical representations of design concepts, 3D printers help in assessing the feasibility of different design options and their potential impact on project costs, leading to a more accurate and realistic project brief.

Interactive Design Exploration: 3D models allow clients and stakeholders to physically explore and interact with design concepts, leading to a more engaged and collaborative briefing process.

Increased Client Confidence: Providing clients with tangible models of design options increases their confidence in the proposed solutions, ensuring that the brief accurately reflects their vision and expectations.

Rapid Iteration and Feedback: 3D printing allows for quick iterations of design concepts, enabling the project team to gather feedback and make adjustments to the brief swiftly, ensuring it is comprehensive and aligned with client goals.

Clarification of Spatial Requirements: 3D models help clarify spatial requirements by providing a clear visual representation of the proposed design, allowing for precise adjustments to the brief to meet functional and aesthetic needs.

In Stage 1, 3D printers play a crucial role in developing a well-defined, accurate, and client-aligned brief, ensuring that the project starts on a solid foundation with clear objectives and stakeholder buy-in.

8.2.5.3 Stage 2: Concept Design

Initial design proposals are developed during this stage, translating the project brief into visual and tangible formats.

Fast Design and Production: The agility of 3D printing in prototyping enhances the concept design stage, facilitating a dynamic design process. This rapid prototyping ability allows for quick iterations, improving communication with clients and stakeholders through tangible models.

Rapid Prototyping of Design Concepts: 3D printers enable the quick creation of physical models of design concepts, allowing designers to explore and test different ideas in tangible form, facilitating the development of more refined and innovative solutions.

Enhanced Visualization: Physical models produced by 3D printers provide a clear, three-dimensional representation of the concept design, helping clients and stakeholders better understand the spatial relationships, scale, and aesthetics of the proposed design.

Effective Design Communication: 3D-printed models serve as powerful tools for communicating design intent to clients, stakeholders, and project teams, ensuring that everyone has a shared understanding of the concept design and reducing the potential for misunderstandings.

Iterative Design Development: Designers can quickly produce and refine multiple iterations of the concept design using 3D printing, allowing for a more iterative and responsive design process that adapts to client feedback and evolving project requirements.

Informed Decision-Making: Tangible 3D models provide clients with a more concrete basis for making decisions about the concept design, helping them to visualize the end product and assess whether it aligns with their vision and functional needs.

Cost and Feasibility Assessment: 3D-printed models can help in assessing the feasibility of different design concepts, providing insights into potential construction challenges, material usage, and costs, leading to more informed and practical design choices.

Client and Stakeholder Engagement: Physical models engage clients and stakeholders more effectively than digital renderings alone, fostering a deeper connection with the design and encouraging active participation in the decision-making process.

Integration of Feedback: 3D printers allow for the rapid incorporation of client and stakeholder feedback into the design, enabling the project team to refine the concept design in real-time and ensuring it meets the desired objectives.

Validation of Design Proportions and Scale: 3D models provide an accurate representation of the design's proportions and scale, allowing the design team to validate these aspects early in the process and make necessary adjustments before moving to more detailed design stages.

Support for Design Reviews: During design reviews, 3D-printed models offer a tangible reference point that facilitates discussions and decision-making, ensuring that the concept design is thoroughly vetted and aligned with project goals.

In Stage 2, 3D printers significantly enhance the concept design process by enabling rapid prototyping, improving communication, and fostering collaboration, leading to more informed decisions and a well-developed design that meets client expectations.

8.2.5.4 Stage 3: Spatial Coordination

This stage involves detailed arrangement of building systems and components to ensure they fit together within the physical space.

Print on Demand: Precision and customization through 3D printing resolve spatial conflicts and ensure seamless system integration. The ability to adjust and optimize designs on demand supports efficient spatial coordination, crucial for the detailed planning of MEP systems.

Accurate Spatial Representation: 3D printers create precise physical models that help visualize and assess the spatial relationships between different building components, ensuring that all elements are well-coordinated and fit together as intended.

Improved Design Coordination: Physical models produced by 3D printers facilitate better coordination among architects, engineers, and other consultants by providing a clear, tangible reference for discussing and resolving spatial challenges.

Clash Detection and Resolution: 3D-printed models enable the early identification and resolution of potential clashes between different building systems (e.g., structural, MEP), reducing the risk of costly design changes later in the project.

Enhanced Communication: Tangible 3D models make it easier to communicate complex spatial arrangements to clients, stakeholders, and the project team, ensuring that everyone has a clear understanding of how the design will be constructed.

Validation of Design Proposals: By producing physical models, the design team can validate the spatial coordination of the project, ensuring that all components are accurately aligned and that the design is feasible for construction.

Streamlined Decision-Making: 3D-printed models provide a clear visual aid that facilitates quicker decision-making by allowing clients and stakeholders to see the physical layout of the project and make informed choices about any necessary adjustments.

Iterative Testing and Refinement: Designers can use 3D printing to create multiple iterations of spatial layouts, allowing for the testing and refinement of different coordination strategies, leading to a more efficient and cohesive design.

Support for Construction Planning: Physical models help in planning construction sequences by providing a clear understanding of how different elements will be assembled, reducing the likelihood of construction delays or issues.

Increased Stakeholder Confidence: Providing stakeholders with a tangible model of the coordinated design increases their confidence in the project's progress and alignment with their expectations, reducing the risk of misunderstandings or disputes.

Enhanced Review Process: 3D-printed models serve as a valuable tool during design reviews, enabling a detailed examination of the spatial coordination and facilitating constructive feedback that can be quickly incorporated into the design.

In Stage 3, 3D printers play a crucial role in ensuring that the design is spatially coordinated, well-communicated, and feasible for construction, leading to a more efficient and successful project outcome.

8.2.5.5 Stage 4: Technical Design

Final design decisions are made, and construction information is detailed, including specifications for materials, finishes, and systems.

Waste Minimization: 3D printing significantly reduces material waste, aligning with sustainable design principles. The precision of 3D printing supports a lean construction approach, enhancing material efficiency and reducing the environmental impact of construction.

Detailed Design Visualization: 3D printers allow for the creation of highly detailed models that accurately represent complex technical components, helping to visualize intricate design elements and ensuring they are well understood before construction begins.

Precision in Design Development: Physical models produced by 3D printers provide precise representations of technical details, allowing designers to identify and resolve potential issues with complex assemblies, joints, or connections early in the design process.

Improved Design Communication: 3D-printed models serve as effective tools for communicating technical design details to clients, stakeholders, and construction teams, ensuring everyone has a clear understanding of the design's intricacies.

Enhanced Collaboration Across Disciplines: 3D printing facilitates better collaboration between architects, engineers, and specialist consultants by providing a tangible reference point for discussing and coordinating the technical aspects of the design.

Validation of Constructability: By creating physical prototypes of technical components, 3D printers help to validate the constructability of the design, ensuring that all elements can be manufactured and assembled as intended.

Streamlined Design Reviews: During technical design reviews, 3D-printed models offer a hands-on way to examine and critique design details, leading to more productive discussions and faster resolution of potential issues.

Support for Material and Component Testing: 3D-printed models can be used to test the performance and fit of materials and components, allowing for adjustments to be made before final specifications are issued, reducing the risk of errors during construction.

Efficient Iteration of Technical Details: Designers can quickly iterate and refine technical details using 3D printing, enabling a more agile design process that can adapt to feedback and evolving project requirements.

Increased Accuracy in Construction Documentation: The precise nature of 3D-printed models ensures that construction documentation is based on accurate representations of the design, reducing discrepancies between the design intent and the built outcome.

Enhanced Stakeholder Confidence: Providing clients and stakeholders with detailed 3D models of technical designs increases their confidence in the project, ensuring that they fully understand the design's complexity and are satisfied with the proposed solutions.

In Stage 4, 3D printers play a vital role in refining and communicating the technical design, ensuring that all details are accurately developed, thoroughly vetted, and ready for successful implementation in the construction phase.

8.2.5.6 Stage 5: Manufacturing and Construction

Construction involves manufacturing components and assembling them on-site.

Environmental Benefit: The environmental benefits of 3D printing during construction include minimized waste and the use of sustainable materials. The technology's efficiency reduces the environmental impact associated with construction activities, contributing to greener building practices.

Prototyping and Testing: 3D printers allow for the creation of prototypes of complex components, enabling manufacturers to test and refine elements before full-scale production, reducing the risk of errors during construction.

Customization and Precision: 3D printing enables the production of custom components with high precision, ensuring that parts fit perfectly during assembly and reducing the need for on-site adjustments.

Speed of Production: By using 3D printers, certain construction components can be produced more quickly compared to traditional manufacturing methods, accelerating the overall construction timeline.

Reduction of Material Waste: 3D printing is an additive manufacturing process, which means materials are used more efficiently, reducing waste and contributing to more sustainable construction practices.

Enhanced Quality Control: 3D printers produce components with consistent quality, minimizing variability in construction elements and ensuring that all parts meet the required specifications.

Simplified Complex Geometries: 3D printing allows for the easy creation of complex geometries that would be difficult or impossible to produce with traditional methods, expanding the possibilities for innovative architectural and structural designs.

On-Demand Manufacturing: Components can be printed on demand, reducing the need for large inventories and enabling just-in-time delivery to the construction site, improving logistical efficiency.

Improved Collaboration with Construction Teams: 3D-printed models of construction components provide a tangible reference for construction teams, ensuring they have a clear understanding of how elements should be assembled and reducing the likelihood of construction errors.

Enhanced Visualization of Construction Sequences: 3D-printed models can be used to visualize and plan construction sequences, helping project managers optimize the order of operations and identify potential challenges before they occur on site.

Support for Prefabrication: 3D printing supports the prefabrication of building elements, which can be assembled on-site, reducing construction time, labour costs, and disruptions to the site environment.

In Stage 5, 3D printers contribute to a more efficient, precise, and innovative construction process by enabling rapid prototyping, custom component production, and improved quality control, ultimately leading to better construction outcomes and streamlined project delivery

8.2.5.7 Stage 6: Handover

The building is completed and handed over to the client, with final documentation and resolution of any issues.

Reduced Turnaround Times and Manual Labour: The efficiency of 3D printing facilitates a smoother transition to building operation by meeting last-minute fabrication needs swiftly. This capability ensures that custom fixtures or architectural details can be incorporated without impacting project timelines.

Creation of Accurate As-Built Models: 3D printers can produce detailed as-built models of the completed project, providing a precise physical representation of the building for the client, which can be used for future reference, maintenance, or alterations.

Enhanced Client Communication: Physical 3D models facilitate clear communication during the handover process, allowing clients to better understand the spatial layout, design features, and technical aspects of their new building.

Improved Training for Facility Management: 3D-printed models of key building systems (e.g., HVAC, plumbing, electrical) can be used to train facility management teams, providing them with a hands-on understanding of the building's operational features.

Support for Maintenance Planning: Detailed 3D models can help facility managers visualize and plan maintenance tasks, ensuring that they fully understand how different systems are integrated and how to access them for repairs or upgrades.

Enhanced Documentation: 3D-printed models can serve as a tangible supplement to digital and paper-based documentation, providing an additional layer of clarity and precision in the project handover package.

Increased Client Satisfaction: Providing clients with a physical model of their new building enhances their overall satisfaction with the project, as it allows them to physically engage with the design and better appreciate the final product.

Facilitation of Post-Occupancy Adjustments: If post-handover adjustments are needed, 3D-printed models can help identify and plan these modifications accurately, ensuring that any changes are seamlessly integrated into the existing design.

Improved Stakeholder Engagement: Stakeholders involved in the handover process, such as investors or regulatory bodies, can benefit from viewing a 3D-printed model, which provides a clear and accessible understanding of the project's final outcome.

Enhanced Record Keeping: A 3D-printed model serves as a valuable physical record of the completed project, which can be stored alongside digital records for long-term archival and reference purposes.

Support for Marketing and Promotion: A detailed 3D model of the completed building can be used for marketing and promotional purposes, showcasing the project's design and quality to potential clients, tenants, or the public.

In Stage 6, 3D printers enhance the handover process by providing accurate, tangible models that improve communication, training, documentation, and client satisfaction, ensuring a smooth and effective transition to the building's operational phase.

8.2.5.8 Stage 7: Use

The building's performance is monitored, and maintenance is carried out during this phase to ensure it meets occupants' needs.

Sustainable Method of Delivery: The operational phase benefits from 3D printing through efficient maintenance and adaptability. The technology's ability to fabricate parts on demand supports building maintenance, enhancing the building's sustainability and adaptability over its lifecycle.

Custom Maintenance Solutions: 3D printers can produce customized components or replacement parts on-demand, enabling quick repairs and minimizing downtime for building operations, thus ensuring continuous functionality.

Enhanced Facility Management: Detailed 3D models of building systems can assist facility managers in visualizing and understanding complex layouts, improving their ability to plan and execute maintenance tasks effectively.

Ongoing Modifications and Upgrades: 3D printing allows for the easy production of parts needed for building modifications or upgrades, facilitating seamless integration with existing structures and systems.

Support for User Training: 3D-printed models can be used to train building users or maintenance staff on specific systems or operational procedures, ensuring they are well-prepared to manage the building's features.

Long-Term Record Keeping: Physical 3D models provide a durable, tangible record of the building's as-built condition, which can be invaluable for future renovations, historical documentation, or legal purposes.

Space Optimization: 3D models of the building's layout can help users explore and plan space optimization strategies, allowing for better utilization of available areas based on evolving needs.

Enhanced User Engagement: Occupants and visitors can engage with 3D-printed models to gain a deeper understanding of the building's design and functionality, enhancing their connection to the space.

Sustainability Initiatives: 3D printing can support sustainability initiatives by enabling the creation of parts using recycled materials, contributing to the building's overall environmental performance over its lifecycle.

Facilitation of Post-Occupancy Evaluations: 3D models can be used to conduct detailed post-occupancy evaluations, helping to identify areas for improvement and ensure that the building continues to meet the needs of its users.

Support for Future Renovations: When renovations or extensions are required, 3D-printed models can assist in planning and executing these changes, ensuring they are well-integrated with the existing structure.

In Stage 7, 3D printers play a crucial role in the ongoing use, maintenance, and evolution of the building, providing customized solutions, enhancing facility management, and supporting continuous improvement throughout the building's lifecycle.

8.2.6 Benefits of 3D Laser Scanners

Some of the benefits of 3D laser scanning are discussed in the ensuing paragraphs.

8.2.6.1 *Stage 0: Strategic Definition*

At the Strategic Definition stage, the emphasis is on understanding the client's business case and strategic brief. Here, improved accuracy from 3D laser scanning is pivotal. By capturing the exact conditions of the existing site or structure, stakeholders can make informed decisions from the outset. This foundational accuracy sets the stage for realistic goal setting and strategic planning, minimizing the risk of costly revisions in subsequent stages.

Accurate Site Data Collection: 3D laser scanners provide precise and comprehensive data on existing site conditions, enabling informed decision-making from the very start of the project.

Rapid and Detailed Surveys: These scanners can quickly capture detailed measurements of a site, reducing the time required for surveys and ensuring that strategic decisions are based on accurate, up-to-date information.

Enhanced Feasibility Analysis: The high-resolution data from 3D laser scanners allow for a thorough analysis of site constraints, topography, and existing structures, aiding in the evaluation of project feasibility and potential risks.

Improved Stakeholder Communication: 3D laser scans produce detailed visualizations that can be easily shared with stakeholders, enhancing their understanding of the site and supporting more effective communication during the strategic definition phase.

Early Identification of Site Challenges: By providing a precise 3D model of the site, laser scanners help identify potential challenges such as uneven terrain, structural obstacles, or spatial constraints, allowing these issues to be addressed early in the planning process.

Support for Comparative Analysis: 3D laser scanning data enable the creation of accurate digital models that can be used to compare different strategic options, helping to select the best approach for meeting the client's requirements.

Reduction of Risk: The comprehensive data collected by 3D laser scanners reduce the risk of overlooking critical site details, ensuring that the strategic decisions made at Stage 0 are well-informed and minimize the potential for costly errors later in the project.

Enhanced Collaboration: The detailed site data can be easily shared among project team members, fostering collaboration and ensuring that all parties have a clear and consistent understanding of the site's characteristics.

Foundation for Digital Twins: The accurate 3D data collected can serve as the foundation for creating a digital twin of the site, which can be used throughout the project lifecycle for ongoing analysis, planning, and decision-making.

Informed Budgeting and Resource Allocation: By providing detailed information about the site's current state, 3D laser scans support more accurate budgeting and resource planning, ensuring that the strategic plan aligns with financial and logistical realities.

In Stage 0, 3D laser scanners significantly enhance the strategic definition process by providing precise, detailed site data that support informed decision-making, risk reduction, and effective communication, laying a strong foundation for the project's success.

8.2.6.2 *Stage 1: Preparation and Brief*

During the Preparation and Brief stage, clients' requirements are developed into a detailed project brief. **Immediate information** provided by 3D laser scanning allows for a rapid assessment of the feasibility of these requirements against the actual site conditions. This phase benefits significantly from the detailed and precise data that 3D laser scanning delivers, ensuring that the project brief is both ambitious and achievable, firmly grounded in reality.

Immediate Information: The ability to perform laser scanning at any time can provide accessible information early in the project to aid in brief development and preparation.

Improved Accuracy: Early on, accurate models from laser scanners support the establishment of a reliable brief by reducing errors and uncertainties.

Accurate Site Information for Brief Development: 3D laser scanners provide detailed and precise measurements of the existing site, ensuring that the project brief is based on accurate data, which leads to more informed design decisions.

Comprehensive Existing Conditions Analysis: The high-resolution data captured by 3D laser scanners allow for a thorough analysis of existing site conditions, including topography, structures, and utilities, helping to shape a realistic and achievable project brief.

Enhanced Feasibility Studies: Detailed 3D scans enable the creation of accurate models of the site, which can be used to conduct feasibility studies that consider spatial constraints, existing structures, and potential site challenges early in the project.

Effective Stakeholder Communication: 3D laser scans produce visualizations that can be easily shared with clients and stakeholders, improving their understanding of the site and facilitating clearer communication during the briefing process.

Informed Decision-Making: The precise data from 3D laser scans support better decision-making regarding site utilization, design approaches, and potential constraints, ensuring that the project brief reflects the true capabilities of the site.

Early Detection of Potential Issues: By capturing detailed site information, 3D laser scanners help identify potential issues, such as structural deficiencies or spatial conflicts, that need to be addressed in the project brief, reducing the risk of unforeseen challenges later.

Support for Accurate Budgeting: The detailed site data provided by 3D laser scanners help in developing a more accurate budget by allowing for precise estimates of required resources and potential site preparation costs.

Facilitation of Design Brief Iterations: 3D models generated from laser scans allow for quick iterations and adjustments to the project brief as new insights are gained, ensuring that the brief evolves to meet the project's goals effectively.

Foundation for Design Concepts: The accurate 3D site models generated by laser scanners serve as a reliable foundation for developing initial design concepts that align with the project brief and site conditions.

Efficient Collaboration Among Project Teams: The detailed data from 3D laser scans can be easily shared among architects, engineers, and consultants, fostering collaboration and ensuring that all team members have a consistent understanding of the site's parameters.

In Stage 1, 3D laser scanners provide critical support for developing a precise, informed, and well-communicated project brief, ensuring that the project begins with a clear understanding of the site's capabilities and constraints.

8.2.6.3 Stage 2: Concept Design

As the project moves into the Concept Design stage, reduce schedule and cost become crucial benefits. The quick and accurate data collection facilitated by 3D laser scanning enables design teams to explore and iterate design concepts with confidence. The efficiency gained in this stage can lead to significant time and cost savings, as the need for rework due to inaccuracies is greatly reduced (Kim et al., 2024).

Improved Accuracy: At this stage, high accuracy helps to reduce design errors and redo of plans, improving the overall quality of the concept design.

Accurate Site Context Integration: 3D laser scanners provide precise, detailed models of the existing site and its surroundings, enabling architects to seamlessly integrate the concept design into the actual site context, ensuring a harmonious fit between the new design and existing structures.

Enhanced Design Accuracy: The high-resolution data from 3D laser scans ensure that the concept design is based on accurate site dimensions and topography, reducing the likelihood of errors and ensuring that the design is both feasible and realistic.

Improved Visualization for Stakeholders: Detailed 3D models generated from laser scans can be used to create realistic visualizations of the concept design within its actual environment, helping stakeholders better understand and engage with the proposed design.

Streamlined Design Iteration: 3D laser scanning data allow for quick and accurate iterations of the concept design, enabling architects to test and refine different design ideas while maintaining alignment with the site's constraints and opportunities.

Early Detection of Design Challenges: By providing a comprehensive view of the site's existing conditions, 3D laser scanners help identify potential design challenges early in the concept stage, such as spatial conflicts or structural limitations, allowing for proactive adjustments.

Support for Collaborative Design: The detailed site data from 3D laser scans can be shared with the entire design team, including engineers and consultants, ensuring that all disciplines work with the same accurate information, leading to a more cohesive and integrated concept design.

Optimized Use of Space: 3D laser scanning provides precise measurements that help in optimizing the use of available space, allowing designers to maximize the functionality and aesthetic appeal of the concept design.

Informed Decision-Making: The accurate data from 3D laser scans support informed decision-making by providing a clear understanding of the site's constraints and possibilities, ensuring that the concept design is both innovative and practical.

Foundation for Sustainable Design: 3D laser scans enable the accurate analysis of site conditions, which can be used to inform sustainable design strategies, such as optimizing natural light, ventilation, and energy efficiency within the concept design.

Enhanced Client Confidence: Providing clients with realistic, data-driven visualizations and models based on 3D laser scans increases their confidence in the concept design, helping to secure buy-in and approval at an early stage.

In Stage 2, 3D laser scanners significantly enhance the concept design process by providing accurate site data, improving visualization, facilitating collaboration, and ensuring that the design is both feasible and aligned with the project's goals.

8.2.6.4 Stage 3: Developed Design

The Developed Design stage benefits from the elimination of the need for manual labour. The precision of 3D laser scanning minimizes human errors in translating site conditions into design developments. This accuracy is crucial for refining the design with detailed technical specifications, ensuring that the project can progress smoothly into construction without the need for costly and time-consuming corrections.

Accurate As-Built Verification: 3D laser scanners provide precise, detailed models of existing structures and site conditions, enabling the design team to verify that the developed design accurately reflects the as-built environment and integrates seamlessly with existing elements.

Improved Spatial Coordination: The detailed data from 3D laser scans facilitate the coordination of architectural, structural, and MEP systems, ensuring that all components fit together without conflicts, reducing the likelihood of costly revisions later in the project.

Enhanced Design Precision: With the high-resolution data from 3D laser scans, designers can develop more precise and detailed plans, ensuring that every aspect of the design is fully resolved and ready for technical detailing in the next stage.

Streamlined Design Validation: 3D laser scans allow for quick validation of the developed design against existing site conditions, helping to ensure that all proposed changes are feasible and do not negatively impact the site or surrounding structures.

Effective Stakeholder Communication: The accurate, detailed models generated from 3D laser scans can be used to create realistic visualizations of the developed design, improving communication with stakeholders and facilitating more informed decision-making.

Reduced Risk of Design Errors: By providing a comprehensive, accurate representation of the site, 3D laser scanners help identify and resolve potential design errors before they become significant issues, reducing the risk of delays and cost overruns during construction.

Support for Detailed Design Adjustments: As the design progresses, 3D laser scanning data can be used to make precise adjustments to the developed design, ensuring that it continues to meet the project's goals and responds to any changes in site conditions or client requirements.

Optimized Use of Space and Resources: The detailed spatial data provided by 3D laser scanners enable designers to optimize the use of available space and resources, leading to more efficient and effective design solutions.

Foundation for Technical Detailing: The accurate models produced by 3D laser scans serve as a reliable foundation for the technical design phase, ensuring that all dimensions, alignments, and spatial relationships are clearly defined and ready for detailed documentation.

Increased Client Confidence: Providing clients with highly accurate, data-driven models based on 3D laser scans increases their confidence in the developed design, helping to secure approval and maintain project momentum.

Reduction in the Number of Site Visits: Fewer site visits during design development allow designers to rely on accurate, detailed scans, reducing unnecessary trips.

Eliminate Needs for Manual Labour: Reducing human errors during design by replacing manual methods with 3D laser scanning ensures more precise development of the design.

Improved Accuracy: Continued benefit in refining and developing the design with more precision

In Stage 3, 3D laser scanners play a crucial role in refining and validating the developed design, ensuring precision, reducing risks, and enhancing communication and collaboration across the project team.

8.2.6.5 Stage 4: Technical Design

At the Technical Design stage, the project benefits from reduction in the number of site visits. The comprehensive data captured by 3D laser scanning mean that much of the work in finalizing construction details can be done remotely, reducing the need for physical site visits. This not only saves time but also reduces the carbon footprint associated with travel to and from the site, contributing to more sustainable project development practices.

Reduce Schedule and Cost: 3D laser scanning reduces time and cost by providing highly accurate data in shorter timespans during detailed technical designs.

Improved Accuracy: The high accuracy of the models plays a critical role in technical detailing, minimizing the need for revisions.

Precise Technical Detailing: 3D laser scanners provide highly accurate site measurements and data, allowing for precise detailing of technical elements, ensuring that all components and systems fit together perfectly in the final construction.

Improved Design Coordination: The detailed 3D models generated from laser scans facilitate the coordination between architectural, structural, and MEP systems, helping to identify and resolve potential clashes or spatial conflicts before construction begins.

Enhanced Accuracy in Construction Drawings: Laser scanning data ensure that all technical drawings and specifications are based on accurate, real-world site conditions, reducing the likelihood of errors and discrepancies during construction.

Streamlined Prefabrication Process: The precision provided by 3D laser scans supports the prefabrication of building components, ensuring that they are manufactured to exact specifications and fit seamlessly on-site, reducing construction time and waste.

Risk Reduction: By providing a detailed, accurate representation of the site, 3D laser scanners help to minimize risks associated with construction errors, delays, and unforeseen site conditions, leading to a smoother construction process.

Enhanced Quality Assurance: The detailed models from 3D laser scans enable thorough quality checks and validation of the technical design against site realities, ensuring that the design is buildable and meets all required standards.

Efficient Design Revisions: If design changes are needed, 3D laser scanning data allow for quick and accurate revisions, ensuring that adjustments are integrated smoothly and do not disrupt the overall project timeline.

Support for Detailed Construction Planning: The precise data from 3D laser scans help in developing detailed construction plans and sequences, improving the efficiency and effectiveness of the construction process.

Improved Communication with Contractors: 3D models based on laser scans provide a clear and accurate reference for contractors, ensuring that they fully understand the technical requirements and can execute the design with confidence.

Accurate As-Built Documentation: As the project progresses, 3D laser scanning can be used to update the as-built documentation, ensuring that it accurately reflects the completed construction and can be used for future maintenance or renovations.

In Stage 4, 3D laser scanners significantly enhance the technical design process by ensuring precision, improving coordination, reducing risks, and supporting the creation of accurate construction documentation, leading to a more efficient and successful project execution.

8.2.6.6 Stage 5: Manufacturing and Construction

During Construction, reduction in the number of site disruptions becomes a key advantage. The detailed planning enabled by 3D laser scanning data allows for a smoother construction process with fewer interruptions. Teams can work more efficiently, with a clear understanding of the site conditions and requirements, leading to a streamlined construction phase with minimized delays and conflicts.

Reduce Schedule and Cost: Time and cost savings due to fewer delays and less rework during the construction process can help streamline manufacturing and construction phases.

Reduction in the Number of Site Disruptions: Reduced site visits mean fewer disruptions to ongoing construction activities, which is key in this phase.

Improve Health and Safety: Less travel to and from site improves health and safety by minimizing the risk of accidents for surveyors during construction.

Accurate Site and Structure Verification: 3D laser scanners provide precise, real-time data on the site and existing structures, ensuring that manufactured components align perfectly with on-site conditions, reducing the risk of fitment issues during construction.

Enhanced Quality Control: Laser scanning enables continuous monitoring of construction progress, allowing for the early detection of deviations from the design, ensuring that all work meets the required specifications and quality standards.

Efficient As-Built Documentation: As construction progresses, 3D laser scans can be used to create accurate as-built models, providing a detailed record of the construction process and verifying that the build conforms to the design.

Streamlined Prefabrication Process: The precise data from 3D laser scanners ensure that prefabricated components are manufactured to exact specifications, reducing assembly time on-site and minimizing the need for modifications.

Improved Construction Accuracy: By providing detailed, up-to-date site measurements, 3D laser scanning ensures that construction teams have the most accurate data, leading to higher accuracy in the placement and **installation** of building elements.

Reduced Rework and Waste: Early detection of discrepancies through laser scanning reduces the likelihood of costly rework and material waste, contributing to a more efficient and sustainable construction process.

Enhanced Collaboration and Communication: The detailed 3D models produced by laser scanners facilitate better communication between architects, engineers, contractors, and other stakeholders, ensuring that everyone is aligned and working with the same accurate information.

Support for Complex Geometries: 3D laser scanners enable the precise construction of complex geometries that might be difficult to achieve with traditional measurement methods, expanding the possibilities for innovative architectural designs.

Real-Time Progress Monitoring: Laser scanning allows for real-time monitoring of construction progress, enabling project managers to compare ongoing work against the design model and make informed decisions quickly.

Improved Safety on Site: By providing accurate data about the construction site, including potential hazards or spatial constraints, 3D laser scanners contribute to creating a safer working environment for construction teams.

In Stage 5, 3D laser scanners play a crucial role in enhancing the accuracy, efficiency, and safety of the manufacturing and construction process, leading to higher-quality outcomes, reduced waste, and improved collaboration across all project stakeholders.

8.2.6.7 Stage 6: Handover and Close-Out

During the Handover and Close-Out stage, enhancing health and safety emerges as a key benefit. The precise and comprehensive data obtained through 3D laser scanning throughout the project lifecycle play a pivotal role in minimizing risks during this critical phase. By providing detailed records of the construction process, informed by accurate scanning data, stakeholders can ensure that the final building adheres to all safety and performance standards. This not only protects the end-users but also safeguards the professionals involved in the project delivery.

Accurate As-Built Documentation: 3D laser scanners create precise as-built models of the completed project, providing clients with detailed and accurate records that reflect the final construction, which are invaluable for future maintenance, renovations, or asset management.

Efficient Verification of Project Completion: Laser scanning enables a thorough comparison between the completed construction and the original design, ensuring that all elements have been built according to specifications before the final handover.

Enhanced Quality Assurance: By capturing every detail of the finished project, 3D laser scanners help verify that the construction meets the required quality standards, identifying any discrepancies or defects that need to be addressed during the close-out process.

Streamlined Handover Process: The detailed models and documentation produced by 3D laser scanners simplify the handover process by providing clear, comprehensive information that can be easily shared with clients, facility managers, and other stakeholders.

Improved Client Communication: 3D models generated from laser scans offer a clear and accurate visual representation of the final project, improving communication with the client and ensuring they fully understand the delivered product.

Support for Maintenance and Operation: The precise as-built data collected by 3D laser scanners can be used to inform and optimize future maintenance and operation plans, ensuring that the building performs as intended throughout its lifecycle.

Facilitation of Post-Occupancy Evaluation: Accurate as-built models enable a thorough post-occupancy evaluation by providing a reliable reference for assessing how the building functions in use and identifying areas for improvement.

Increased Transparency and Trust: Providing clients with detailed, accurate as-built documentation builds trust and ensures transparency, as the client can see that the project has been completed to the agreed-upon specifications.

Foundation for Digital Twins: The high-resolution data from 3D laser scans can be used to create a digital twin of the building, which can serve as a dynamic tool for ongoing building management, monitoring, and future upgrades.

Legal and Contractual Compliance: The precise documentation provided by 3D laser scanners can help ensure that all contractual obligations have been met, reducing the risk of disputes and facilitating the close-out of contracts.

In Stage 6, 3D laser scanners enhance the handover and close-out process by providing accurate as-built documentation, improving quality assurance, and supporting future building management and maintenance, leading to a smooth and successful project conclusion.

8.2.6.8 Stage 7: Use

Finally, during the Use stage, the operational phase of the building, the improved accuracy, and immediate information provided by 3D laser scanning can greatly facilitate maintenance, renovation, and adaptation activities. The detailed as-built models created with laser scanning serve as a valuable resource for facilities management teams, helping to ensure the building continues to meet users' needs efficiently and sustainably over time. Accurate models enable better planning for future changes and can significantly reduce the costs and disruptions associated with retrofitting or extending the building (Chang et al., 2024).

Reduction in the Number of Site Visits: Fewer site visits during operation and maintenance due to reliance on accurate 3D models reduce ongoing disruptions and costs.

Accurate Facility Management: 3D laser scanners provide precise as-built models that can be used for effective facility management, allowing for accurate maintenance planning and space management throughout the building's lifecycle.

Streamlined Maintenance and Repairs: The detailed 3D models help identify and locate building components accurately, making it easier to plan and execute maintenance tasks or repairs with minimal disruption to operations.

Support for Renovations and Upgrades: When renovations or upgrades are needed, 3D laser scans provide accurate data on existing conditions, facilitating seamless integration of new elements with the existing structure and reducing the risk of design conflicts.

Optimized Space Utilization: Facility managers can use 3D laser scanning data to analyse and optimize the use of space, ensuring that the building's layout continues to meet the evolving needs of its occupants.

Enhanced Safety Management: The detailed models generated by 3D laser scanners help in monitoring and managing safety aspects within the building, such as fire exits, escape routes, and potential hazards, ensuring compliance with safety regulations.

Effective Asset Management: 3D laser scanning enables accurate tracking and management of building assets, ensuring that all components are accounted for and maintained according to the necessary standards.

Facilitation of Building Performance Monitoring: The accurate data from 3D laser scans can be used to monitor building performance over time, helping to identify areas where improvements can be made to enhance energy efficiency and occupant comfort.

Support for Digital Twin Development: 3D laser scanning data can be used to create and maintain a digital twin of the building, providing a dynamic and up-to-date model that supports ongoing building management, monitoring, and future planning.

Accurate Documentation for Compliance: The detailed as-built documentation provided by 3D laser scanners ensures that all building records are accurate and up-to-date, supporting compliance with regulatory requirements and facilitating audits or inspections.

Long-Term Cost Savings: By providing accurate data for maintenance, repairs, and upgrades, 3D laser scanning helps reduce the overall cost of building management over time, as tasks can be planned more effectively and executed with greater precision.

In Stage 7, 3D laser scanners play a crucial role in the ongoing use and management of the building, enhancing facility management, supporting renovations, and ensuring that the building continues to operate efficiently and effectively throughout its lifecycle.

8.3 Conclusions

In crafting a detailed conclusion on the transformative impact of digital construction technologies, we delve into the collective advancements brought forth by BIM, robotics, drones, 3D laser scanners, 3D printers, and VR/AR/MR technologies. These innovations represent a significant leap forward for the construction industry, offering unprecedented levels of efficiency, accuracy, and safety.

BIM stands at the forefront of this revolution, facilitating a collaborative digital environment that enhances visualization, reduces errors, and streamlines project management. Robotics follow, introducing automation to repetitive and hazardous tasks, thereby not only improving safety but also compensating for labour shortages. Drones offer aerial perspectives for site surveys and inspections, delivering real-time data and accessing previously unreachable areas. 3D laser scanners contribute by capturing precise measurements quickly, improving as-built documentation quality. 3D printing introduces a new era of material fabrication, allowing for complex designs while potentially reducing waste and costs. Finally, VR/AR/MR technologies enrich visualization for clients and stakeholders, offer immersive training environments, and support the design and construction processes.

However, the adoption of these technologies is not without challenges, including the need for significant initial investment, specialized training, and the integration of new tools into existing workflows. Despite these hurdles, the construction industry is witnessing a paradigm shift towards more sustainable, efficient, and innovative practices, driven by digital technologies.

The future promises even greater integration of AI and machine learning, further enhancing the capabilities of digital construction technologies. As the industry continues to evolve, it will increasingly rely on these technological advancements to meet the demands of modern construction projects, emphasizing the importance of continuous research, development, and training. This digital transformation is not merely a trend but a fundamental shift towards achieving greater efficiency, innovation, and sustainability in construction. The call to action for industry professionals is clear: embrace and invest in digital technologies to unlock their full potential, shaping the future of construction in the process.

Chapter 9

Barriers to Digital Construction Technologies Adoption

9.1 Introduction

In this section, we will begin by discussing the overarching barriers that hinder the adoption of digital construction technologies, providing a broader context for understanding the challenges faced by the industry. Following this, we will delve into the specific barriers associated with key digital technologies, including BIM, robotics, drones, augmented, virtual, and mixed reality (AR/VR/MR), 3D printing, and 3D laser scanning. Each technology presents unique challenges, and understanding these will be critical to evaluating their integration and future potential in construction practices.

9.2 Barriers to the Adoption of Digital Construction Technologies

Barriers to the adoption of digital construction technologies in the construction industry hinder the widespread implementation of innovative and advanced technological solutions. Despite the potential benefits and opportunities that digital transformation can offer to the construction sector, several challenges and resistance factors have impeded its full integration. Some of the key barriers include:

9.2.1 Barriers to BIM Adoption

9.2.1.1 Stage 0: Strategic Definition

Lack of Early-Stage Data: BIM relies on detailed information, but at the strategic definition stage, there may be limited data available, making it challenging to fully utilize BIM's capabilities for decision-making and analysis.

High Initial Costs: Implementing BIM requires investment in software, hardware, and training, which can be a significant barrier at the early stages of a project when budgets are still being defined.

Limited Stakeholder Engagement: Not all stakeholders may be familiar with BIM at the strategic level, leading to potential resistance or misunderstandings about its value and capabilities, which can hinder collaboration and buy-in.

Difficulty in Defining Clear Objectives: The strategic definition phase is focused on setting broad project goals, and the abstract nature of this stage may not align well with BIM's structured and data-driven approach, making it difficult to set clear BIM objectives.

DOI: 10.1201/9781003476740-9

Complexity of Integrating BIM with Strategic Planning: Integrating BIM into strategic planning requires aligning it with business goals and project outcomes, which can be complex and time-consuming, especially if the organization lacks BIM expertise.

Uncertain ROI at Early Stages: The return on investment (ROI) for BIM can be unclear during the strategic definition phase, as the benefits of BIM typically become more evident in later stages of the project, leading to hesitancy in committing to its use.

Inconsistent Standards and Practices: The lack of standardized practices and protocols for BIM at the strategic definition stage can result in inconsistencies and inefficiencies, particularly in projects that involve multiple stakeholders or cross-disciplinary teams.

Challenges in Aligning BIM with Business Case Development: Developing a business case for the project can be challenging when trying to incorporate BIM, as it requires a clear understanding of how BIM will add value to the strategic goals and outcomes, which may not be fully defined yet.

Limited Use of BIM for High-Level Decision-Making: BIM is often more effective at detailed design and construction stages, while its application in high-level, strategic decision-making is less established, potentially limiting its impact at this phase.

In the Strategic Definition phase, these barriers and challenges can hinder the effective integration and utilization of BIM, requiring careful consideration and planning to overcome them and realize the potential benefits of BIM in later stages of the project.

9.2.1.2 Stage 1: Preparation and Brief

Inconsistent Information Requirements: At the preparation and brief stage, the specific information required for BIM can be unclear or inconsistent, leading to challenges in setting up the BIM environment effectively.

Stakeholder Alignment Issues: Not all stakeholders may be familiar with or committed to using BIM, leading to difficulties in achieving consensus on BIM objectives, protocols, and deliverables during the briefing process.

High Initial Setup Costs: The early implementation of BIM requires investment in technology and training, which can be a barrier when project budgets are still being defined and resources are limited.

Complexity in Defining BIM Execution Plans: Developing a detailed BIM Execution Plan (BEP) during the preparation and brief stage can be complex and time-consuming, particularly when the project scope and requirements are still evolving.

Difficulty in Capturing Client Requirements: Translating client requirements into BIM-compatible data and models can be challenging, especially when clients are unfamiliar with BIM or their needs are not yet fully articulated.

Limited Experience and Expertise: The preparation and brief phase often involves high-level decision-making, and the project team may lack the necessary BIM expertise to effectively integrate it into the early project stages.

Resistance to Change: Introducing BIM at this stage may encounter resistance from stakeholders who are accustomed to traditional processes and are hesitant to adopt new technologies and workflows.

Challenges in Data Integration: Integrating data from various sources into a cohesive BIM model can be difficult, particularly when dealing with legacy systems or incomplete information at the early stages of the project.

Time Constraints: The preparation and brief phase is often under tight time constraints, making it challenging to allocate sufficient time for thorough BIM setup and planning, which could impact the quality of the BIM process later on.

Uncertainty in Scope and Deliverables: The scope and deliverables of the project may not be fully defined during this stage, making it difficult to establish clear BIM goals and outputs, which could lead to inefficiencies and rework as the project progresses.

In the Preparation and Brief stage, these barriers and challenges can complicate the effective implementation of BIM, requiring careful planning, stakeholder engagement, and clear communication to ensure that BIM can be successfully integrated into the project as it moves forward.

9.2.1.3 Stage 2: Concept Design

Limited Flexibility for Creative Design: BIM's structured and data-driven approach can sometimes constrain the creative exploration of design concepts, limiting the ability to experiment with innovative or unconventional ideas during the concept design phase.

High Learning Curve: Designers may face a steep learning curve when using BIM software for concept design, particularly if they are more accustomed to traditional design tools, potentially slowing down the design process.

Time-Intensive Model Setup: Setting up a BIM model in the early concept design phase can be time-consuming, especially when project details are still evolving, leading to delays in the design process.

Integration Challenges with Other Design Tools: Designers often use a variety of tools for concept design (e.g., sketching, 3D modelling software) that may not easily integrate with BIM, resulting in workflow inefficiencies and potential data loss.

Difficulty in Managing Design Iterations: BIM's focus on accuracy and detail can make it challenging to quickly iterate and explore multiple design options, which is a key aspect of the concept design phase.

High Initial Cost and Resource Demand: The implementation of BIM during the concept design stage requires significant investment in software, hardware, and training, which may be seen as a barrier when the design is still in its formative stages.

Complexity in Setting Up Design Standards: Establishing BIM standards and protocols that align with the project's concept design goals can be complex and time-consuming, especially when working with multiple stakeholders or interdisciplinary teams.

Difficulty in Conveying Early Design Intent: The level of detail required by BIM can make it challenging to communicate broad design ideas and concepts, which may be better conveyed through simpler, more abstract visualizations at this stage.

Limited Client Engagement: Clients may struggle to engage with the highly detailed and technical nature of BIM models during the concept design phase, potentially leading to misunderstandings or misalignment with their expectations.

Data Overload: The extensive data and detail required by BIM can overwhelm the design team during the concept phase, where the focus should be on exploring and refining broad design ideas rather than detailed technical solutions.

In the Concept Design stage, these barriers and challenges can hinder the fluidity and creativity of the design process, requiring careful management to balance the use of BIM with the need for flexibility and rapid exploration of design ideas.

9.2.1.4 *Stage 3: Spatial Coordination*

Complexity of Integrating Multiple Disciplines: BIM requires the integration of models from various disciplines (architectural, structural, MEP, etc.), which can be complex and challenging, particularly if teams are using different software platforms or standards.

Data Management Challenges: Managing and coordinating large volumes of data from different disciplines can be overwhelming, leading to potential data conflicts, version control issues, and increased risk of errors in the coordinated model.

Interoperability Issues: Despite advancements in BIM software, interoperability between different BIM tools and platforms remains a challenge, potentially leading to difficulties in integrating models and ensuring seamless collaboration across teams.

High Demand for Accuracy: Spatial coordination in BIM demands a high level of accuracy and precision, which can be time-consuming to achieve and may require significant effort to maintain throughout the project, especially with frequent design changes.

Steep Learning Curve for Coordination Tools: The tools used for BIM coordination, such as clash detection software, often have a steep learning curve, which can pose a challenge for team members who are not familiar with these tools, potentially slowing down the coordination process.

Clash Detection Overload: While BIM enables automated clash detection, the process can generate a large number of minor or irrelevant clashes, overwhelming the team and making it difficult to prioritize and address the most critical issues.

Time-Consuming Model Updates: Keeping the coordinated BIM model up to date with design changes from various disciplines can be time-consuming, particularly when changes are frequent or occur late in the design process, leading to potential delays.

Communication Gaps Between Teams: Effective spatial coordination requires continuous communication between all design teams, but miscommunication or lack of collaboration can lead to coordination errors and unresolved clashes in the BIM model.

Resource-Intensive Process: The spatial coordination phase requires significant resources in terms of time, personnel, and technology, which can strain project budgets and timelines, particularly on complex projects.

Resistance to Process Changes: Implementing a coordinated BIM approach often requires changes to established workflows and processes, which can meet resistance from team members accustomed to traditional coordination methods, potentially hindering the adoption of BIM practices.

In the Spatial Coordination stage, these barriers and challenges can complicate the process of integrating and managing multidisciplinary models, requiring careful planning, communication, and resource allocation to ensure effective coordination and minimize potential issues.

9.2.1.5 *Stage 4: Technical Design*

High Level of Detail Required: BIM demands a significant amount of detailed information during the technical design phase, which can be time-consuming to produce and manage, potentially leading to delays in the project timeline.

Complexity in Model Management: As the BIM model becomes more detailed, managing the increasing complexity can be challenging, particularly in ensuring that all elements are correctly integrated and coordinated across various disciplines.

Interdisciplinary Coordination Challenges: The need for precise coordination between architectural, structural, MEP, and other technical disciplines can lead to challenges in ensuring that all models are aligned, particularly when changes occur frequently.

Potential for Data Overload: The extensive amount of data and detail in the BIM model can overwhelm project teams, making it difficult to focus on the most critical aspects of the design and leading to potential oversights.

Software and Hardware Limitations: The technical design phase often requires advanced BIM tools and high-performance hardware, which can be costly and may not be readily available to all team members, potentially limiting the efficiency of the BIM process.

Steep Learning Curve for Technical Design: Mastering the use of BIM software for technical design requires specialized skills and training, which can be a barrier for team members who are not experienced with these tools, leading to potential delays or errors.

Version Control and Data Integrity Issues: Ensuring that the BIM model remains up-to-date and that all team members are working with the latest version can be challenging, particularly in large projects with multiple contributors, leading to potential inconsistencies and errors.

Integration of Non-Standard Components: Incorporating non-standard or custom components into the BIM model can be difficult, as these elements may not be readily available in BIM libraries and may require custom modelling, adding to the project's complexity.

Communication and Collaboration Barriers: Effective communication and collaboration are critical in the technical design phase, but differences in technical knowledge or BIM proficiency among team members can hinder effective collaboration and lead to misunderstandings.

Potential for Increased Costs: The detailed nature of BIM in the technical design phase can lead to increased costs due to the need for specialized software, training, and resources, which may strain project budgets, especially if not anticipated early on.

In the Technical Design stage, these barriers and challenges can complicate the process of developing detailed and accurate designs within a BIM framework, requiring careful management, resource allocation, and collaboration to ensure successful outcomes.

9.2.1.6 Stage 5: Manufacturing and Construction

Integration with Traditional Construction Practices: BIM processes may not align seamlessly with traditional construction methods, leading to challenges in integrating digital models with on-site practices and potentially causing disruptions or inefficiencies.

Data Transfer Issues: Transferring BIM data to subcontractors and manufacturers who may use different software platforms or lack BIM capabilities can lead to data compatibility issues, errors, or loss of information.

On-Site BIM Utilization Challenges: Construction teams on-site may lack the necessary training or equipment to effectively use BIM models, reducing the potential benefits of BIM in guiding the construction process and leading to reliance on traditional methods.

Increased Coordination Requirements: The use of BIM during construction requires continuous coordination and communication between design and construction teams, which can be challenging to maintain, especially on large or complex projects.

Potential for Information Overload: The detailed information provided by BIM models can overwhelm construction teams, making it difficult to focus on the most relevant data for specific tasks and potentially leading to confusion or mistakes.

Cost and Resource Demands: Implementing BIM during construction may require additional resources, such as specialized hardware, software, and training, which can increase project costs and strain budgets, particularly if not planned for from the outset.

Real-Time Model Updates and Revisions: Keeping the BIM model updated with real-time construction progress and changes can be challenging, especially in fast-paced construction environments, leading to potential discrepancies between the model and actual site conditions.

Dependence on Digital Infrastructure: Effective use of BIM on-site requires robust digital infrastructure, including reliable internet access and adequate hardware, which may not always be available, especially in remote or challenging locations.

Resistance to Change: Construction teams may resist adopting BIM due to a preference for traditional methods or a lack of familiarity with digital tools, hindering the full integration of BIM into the construction process.

Complexity in Prefabrication and Modular Construction: While BIM supports prefabrication and modular construction, the complexity of coordinating these processes with on-site activities can be challenging, requiring precise planning and communication to avoid errors and delays.

In the Manufacturing and Construction stage, these barriers and challenges can limit the effectiveness of BIM, requiring careful planning, training, and coordination to fully realize the benefits of BIM in improving construction efficiency and accuracy.

9.2.1.7 *Stage 6: Handover*

Complexity in As-Built Documentation: Creating accurate as-built BIM models that reflect the final construction can be complex and time-consuming, particularly if there were many changes during construction that need to be meticulously documented.

Inconsistent Data Formats: During handover, different stakeholders may use various software platforms, leading to challenges in ensuring that the BIM data is compatible and easily accessible for all parties, including facility managers.

Inadequate Training for Facility Managers: Facility managers may lack the necessary training to effectively use and navigate the BIM model, reducing the value of the detailed data provided at handover and potentially leading to underutilization.

Data Overload: The sheer volume of information contained in a BIM model can overwhelm facility managers, making it difficult to extract the most relevant data needed for operations and maintenance.

Integration with Existing Systems: Integrating the BIM model with existing facility management systems and tools can be challenging, particularly if those systems are not BIM-compatible, leading to potential inefficiencies and data gaps.

Updating and Maintaining the BIM Model: Ensuring that the BIM model remains a living document that is regularly updated post-handover can be challenging, requiring a commitment of resources that may not always be available or prioritized by the building owner.

Transfer of Responsibility: The transition from the construction team to the facility management team involves transferring responsibility for the BIM model, which can lead to issues if the handover process is not well-coordinated or if there is a lack of understanding of the model's value.

Potential Legal and Contractual Issues: The handover phase may raise legal or contractual issues related to the completeness and accuracy of the BIM model, especially if discrepancies between the model and the actual construction are discovered.

Costs Associated with Handover: The process of preparing the BIM model for handover, including ensuring all data is accurate and up-to-date, can incur additional costs, which may strain project budgets if not anticipated.

Resistance to Digital Transition: Stakeholders who are more accustomed to traditional paper-based documentation may resist the transition to digital BIM models, leading to potential conflicts or reluctance to fully adopt BIM practices in facility management.

In the Handover stage, these barriers and challenges can hinder the effective transition of the BIM model to the facility management team, requiring careful planning, training, and coordination to ensure that the full value of BIM is realized in the building's operational phase.

9.2.1.8 Stage 7: Use

Lack of Ongoing Model Updates: Keeping the BIM model updated during the use phase can be challenging, especially if there is no clear process or responsibility for incorporating changes, leading to the model becoming outdated and less useful over time.

Underutilization of BIM Data: Facility managers and building operators may not fully utilize the rich data available in the BIM model due to a lack of training or familiarity with BIM tools, resulting in missed opportunities for optimizing building performance.

Integration with Facility Management Systems: Integrating the BIM model with existing facility management systems can be complex, especially if those systems are not designed to work with BIM data, leading to inefficiencies and potential data silos.

High Costs of Maintenance: Maintaining and updating the BIM model during the building's operational life can be costly, particularly if significant modifications or renovations occur that require extensive updates to the model.

Data Security and Privacy Concerns: The detailed data contained in a BIM model can raise security and privacy concerns, particularly if the model includes sensitive information about building systems or occupant data, requiring robust cybersecurity measures.

Complexity of Data Extraction: Extracting specific, actionable information from a complex BIM model can be difficult, particularly for facility managers who may not have advanced BIM expertise, potentially leading to underuse of the model's capabilities.

Dependence on Digital Infrastructure: The effective use of BIM during the building's operational phase depends on reliable digital infrastructure, including hardware and software, which may require ongoing investment and maintenance.

Resistance to Adoption: Building operators and maintenance staff who are more accustomed to traditional management practices may resist adopting BIM-based processes, leading to a reliance on legacy methods and underutilization of the BIM model.

Inconsistent Data Standards: If the BIM model was not developed with consistent data standards, it might be challenging to use the model effectively for facility management, leading to inefficiencies and potential errors in building operations.

Legal and Compliance Challenges: Ensuring that the BIM model is used in compliance with legal and regulatory requirements can be complex, particularly if there are discrepancies between the as-built model and actual building conditions.

In the Use phase, these barriers and challenges can limit the effectiveness of BIM in supporting building operations and maintenance, requiring ongoing commitment to training, model updates, and integration with facility management practices to fully realize the benefits of BIM throughout the building's lifecycle.

Organizational Readiness: Implementing BIM requires a significant organizational shift in the construction industry. It involves adopting new processes, technologies, and collaborative workflows. Organizations may face challenges in terms of adjusting their existing project management systems and integrating BIM into their current practices. Resistance to change and a lack of understanding of the benefits of BIM within the organization can hinder the readiness to embrace this technology fully.

Lack of Investment: The successful implementation of BIM often necessitates investment in hardware, software, and training for the workforce. For some construction companies, especially smaller ones, this initial investment may seem prohibitive, leading to limited organizational readiness. Without the necessary resources and support, the transition to BIM can be slow and challenging.

Client Requests and Requirements: Client demand plays a crucial role in driving the adoption of BIM in construction projects. If clients are not well-informed about the benefits of BIM or do not explicitly request its implementation, construction companies may not see a strong incentive to invest in BIM technology and processes. Lack of demand from clients can hinder the integration of BIM into projects, resulting in missed opportunities to leverage the technology's advantages.

Limited Education and Training: Implementing BIM requires a specific skill set that includes proficiency in BIM software, understanding of data management, and collaborative work processes. A lack of formal training and education in BIM tools and methodologies can create a skill gap among the workforce. Construction professionals who are not well-versed in BIM might struggle to use the technology effectively, leading to inefficiencies and potential errors in project delivery.

Presently, the extent to which BIM has been embedded in schools and higher education curriculum is very limited. Perhaps partly due to lack of education and training in BIM, lack of expertise in the construction industry is too common in Africa. In Nigeria, lack of expertise within organizations and within project team has been reported as one of the major barriers to the adoption of BIM (Hamma-Adama and Kouider, 2019).

Lack of Standards & Policies: To ensure effective understanding and collection in traditional construction practice, the use of standards is imperative. For example, BS 1192 is often used in guiding collaborative working in traditional working practice. Similarly, standards are required in delivering BIM compliant projects. In the UK, BS1111 are the most recent in delivering BIM compliant projects in the UK. At an international stage, ISO is used in delivery of BIM compliant projects. In Africa, there are no standards that can guide the delivery of projects using BIM.

In many developed countries, policies have been used to drive the BIM adoption. In the UK, the policy to adopt BIM level 2 on all government funded project came into force in 2016. Such policies are lacking for African countries. In Nigeria, the lack of standardization and protocols was found as significant barriers to BIM adoption (Hamma-Adama and Kouider, 2019).

Indigenous Knowledge: Most BIM systems are modelled on construction systems in the global North. For example, the wall, window, and door components used in the design of building models are based on Western style of construction.

Indigenous Knowledge Integration: Collaboratively develop BIM libraries and databases that include Indigenous construction practices, materials, and design elements. This integration will enable Indigenous builders and architects to create BIM models that reflect their unique approaches.

Limited Connectivity: A key requirement to working at BIM level 2 in the UK requires managing building information in a Common Data Environment. Some examples of Common Data Environments are BIM 360, Aconex, and 4Project. These Common Data Environments are cloud-based systems that depend on internet connectivity. Thus, for true collaboration at BIM level 2 to achieved, internet connectivity is imperative. In Olanrewaju et al. (2020), it was argued that the limited or lack of internet connectivity is a major barrier to BIM adoption in Nigeria. Cloud-based BIM platforms enable multiple team members to work on the same model simultaneously. However, limited connectivity can lead to challenges in accessing cloud services, hindering the real-time collaboration and coordination that BIM aims to achieve.

Data Transfer and Synchronization: BIM involves the exchange and synchronization of large amounts of data between various stakeholders, including architects, engineers, contractors, and clients. Limited connectivity, especially in remote or rural areas, can result in slow data transfer, delays in model updates, and difficulties in maintaining real-time collaboration.

Access to Online Resources: BIM often requires access to online databases, libraries, and resources for building components, materials, and manufacturer information. With limited connectivity, accessing these resources becomes difficult, impacting the quality and accuracy of BIM models.

Limited Research Capacities: Research about BIM adoption is generally very low compared to developed countries. A recent study by Saka and Chan (2019) revealed a varying level of BIM growth, with North Africa, West Africa, and Southern Africa leading the research development, whilst East Africa and Central Africa are slightly lagging behind.

Lack of Case Studies and Best Practices: Research plays a crucial role in generating case studies and best practices that showcase successful BIM implementation in various construction projects. Limited research capacities may result in a dearth of real-world examples, making it challenging for industry stakeholders to understand the practical benefits of BIM.

Lack of standardized systems and processes: The lack of standardized systems and processes significantly hinders the adoption of BIM in the construction industry (Alshibani et al., 2024). This issue arises from the diversity of construction projects and the multitude of stakeholders, each employing different BIM software and protocols, complicating interoperability and effective data management throughout the project lifecycle. Such disparities can lead to inconsistencies in information, hinder collaboration, and ultimately, reduce efficiency while increasing project costs. Addressing this barrier necessitates a collective effort towards establishing and embracing industry-wide standards for BIM implementation, ensuring a unified approach that benefits all parties involved in the construction process.

9.2.2 *Barriers to the Adoption Robots*

9.2.2.1 *Stage 0: Strategic Definition*

High Initial Investment: The cost of integrating robots into a project can be substantial, including the purchase, setup, and training, making it difficult to justify their use during the early, strategic stages when budgets are still being defined.

Unclear ROI at Early Stages: The ROI for implementing robotic technology may not be immediately evident in the strategic phase, leading to hesitancy in committing to this technology without clear, long-term benefits.

Lack of Familiarity and Expertise: Project stakeholders may lack experience with robotics, leading to uncertainty and resistance when considering the integration of robots into strategic plans, particularly if the perceived risks outweigh the benefits.

Difficulty in Defining Strategic Value: It can be challenging to define the strategic value that robots will bring to a project at this early stage, particularly when project goals and requirements are still broad and not fully developed.

Integration with Existing Processes: Incorporating robots into existing workflows and processes can be complex, especially when traditional methods are deeply ingrained, leading to potential disruptions or inefficiencies during the strategic planning phase.

Regulatory and Compliance Uncertainty: The use of robots may face regulatory challenges or compliance issues, particularly in industries with strict safety and operational standards, making it difficult to incorporate them into strategic planning without clear guidance.

Limited Flexibility in Early Design Stages: Robots are often designed for specific tasks, which may limit their flexibility in the early stages of a project when design and construction methods are still being explored and defined.

Potential Resistance from Stakeholders: Key stakeholders, including clients and team members, may resist the introduction of robotic technology due to concerns about job displacement, cost, or complexity, hindering its consideration in strategic discussions.

Technology Integration Challenges: Integrating robots with other advanced technologies (e.g., BIM, AI) at the strategic phase can be challenging, particularly if there is a lack of established standards or interoperability between systems.

Uncertain Impact on Project Outcomes: The impact of robots on overall project outcomes may be difficult to predict at the strategic stage, leading to concerns about whether their use will actually contribute to meeting the project's strategic objectives.

In the Strategic Definition phase, these barriers and challenges can limit the consideration and integration of robots into the project, requiring careful evaluation, clear communication, and strategic planning to overcome these obstacles and make informed decisions about their potential use.

9.2.2.2 Stage 1: Preparation and Brief

Complexity in Defining Roles: Determining the specific roles and tasks for robots during the preparation and brief phase can be challenging, especially when project requirements and processes are still being established.

High Initial Costs: The cost of acquiring and implementing robotic systems can be significant, which may be difficult to justify during the early stages of project planning when budgets are typically constrained.

Limited Expertise: The project team may lack the necessary expertise to effectively plan for and integrate robots into the project, leading to potential oversights or unrealistic expectations in the project brief.

Uncertainty in Integration with Existing Processes: Integrating robots with existing workflows and processes can be complex, requiring careful planning to avoid disruptions, inefficiencies, or conflicts with traditional methods.

Resistance to Change: Stakeholders may resist the inclusion of robots in the project due to concerns about job displacement, operational complexity, or unfamiliarity with the technology, hindering their acceptance during the briefing process.

Regulatory and Compliance Challenges: The use of robots may raise regulatory or compliance concerns, particularly in industries with strict safety standards, making it difficult to incorporate them into the project brief without clear guidelines.

Difficulty in Capturing Client Requirements: Translating client requirements into specific robotic applications can be challenging, particularly if the client is unfamiliar with the potential benefits or limitations of robotics in construction.

Limited Flexibility in Early Planning: Robots are often designed for specific tasks, which may limit their flexibility in the early planning stages, making it difficult to adapt them to evolving project needs or design changes.

Unclear Impact on Project Timeline: The potential impact of robots on the project timeline may be uncertain during the preparation and brief phase, leading to difficulties in accurately planning and scheduling their use.

Challenges in Coordinating Multiple Technologies: Integrating robots with other advanced technologies (e.g., BIM, AI) during the brief phase can be challenging, particularly if there is a lack of established protocols or compatibility between systems.

In the Preparation and Brief phase, these barriers and challenges can complicate the effective integration of robots into the project plan, requiring careful consideration, expert input, and clear communication to ensure that their potential benefits are fully realized.

9.2.2.3 Stage 2: Concept Design

Limited Flexibility for Creative Design: Robots are often designed for specific tasks, which may limit their ability to adapt to the fluid and creative nature of the concept design phase, potentially stifling innovation and exploration of design ideas.

Integration with Early-Stage Design Tools: Integrating robotic capabilities with the software and tools commonly used in the concept design phase can be challenging, leading to potential inefficiencies or the need for additional software development.

High Initial Costs: The investment required to incorporate robots into the design process, including hardware, software, and training, can be prohibitive during the early concept design stage when budgets are typically limited.

Complexity in Defining Robotic Applications: It can be difficult to determine the specific roles that robots will play in the concept design phase, particularly when the design is still evolving and the full scope of the project has not yet been defined.

Steep Learning Curve: Designers may face a steep learning curve when incorporating robotics into the concept design process, potentially slowing down the pace of design development and innovation.

Challenges in Visualizing Robotic Implementation: It may be difficult to visualize how robots will be implemented in the final construction process based on early concept designs, leading to uncertainty and potential misalignment between design intent and robotic capabilities.

Potential Resistance from Designers: Designers accustomed to traditional methods may resist the introduction of robotics into the concept design phase, viewing it as an unnecessary complication or a threat to their creative freedom.

Interdisciplinary Coordination Issues: Coordinating the use of robots with other disciplines (e.g., structural engineering, MEP) during the concept design phase can be complex, particularly if there is a lack of clear communication or understanding of robotic capabilities.

Uncertain Impact on Design Outcomes: The impact of robots on the final design outcomes may be uncertain during the concept phase, making it difficult to justify their use without a clear understanding of the potential benefits or risks.

Challenges in Communicating Robotic Integration to Clients: Explaining the role and benefits of robots in the concept design phase to clients can be challenging, particularly if the client is unfamiliar with the technology or concerned about its implications for cost and project complexity.

In the Concept Design phase, these barriers and challenges can hinder the effective integration of robots into the design process, requiring careful planning, clear communication, and a flexible approach to ensure that their potential benefits are fully realized without stifling creativity or innovation.

9.2.2.4 Stage 3: Spatial Coordination

Complexity in Integrating Multiple Systems: Spatial coordination requires the integration of architectural, structural, and MEP systems, which can be challenging when introducing robots, as ensuring compatibility between robotic operations and these systems can be complex.

Limited Adaptability to Design Changes: Robots may have limited flexibility in adapting to design changes that occur during spatial coordination, leading to potential delays or the need for reprogramming and adjustments.

Coordination Between Human and Robotic Workflows: Ensuring seamless coordination between human workers and robots can be difficult, particularly in complex environments where spatial coordination requires precise timing and synchronization of tasks.

High Dependency on Accurate Data: Robots rely on highly accurate spatial data for effective operation. Any inaccuracies or inconsistencies in the BIM model or other data sources can lead to operational errors or misalignments during the coordination phase.

Interoperability Challenges: Integrating robots with various design and coordination software tools can be challenging due to interoperability issues, potentially leading to inefficiencies or the need for custom solutions.

Increased Complexity in Clash Detection: While robots can assist with spatial coordination, they can also add complexity to clash detection processes, as their movements and operational space need to be considered alongside traditional building elements.

Cost Implications: The use of robots in spatial coordination may involve significant costs, including the acquisition of robots, software, and the necessary training, which can strain project budgets, especially in the coordination phase.

Training and Expertise Requirements: Effective use of robots during spatial coordination requires specialized knowledge and skills, which may not be readily available within the project team, leading to potential delays or errors.

Potential Disruption to Established Workflows: Introducing robots into the spatial coordination process may disrupt established workflows, requiring adjustments to traditional methods and potentially causing resistance from team members accustomed to manual coordination.

Reliance on Advanced Technology: The successful use of robots in spatial coordination depends on advanced technology and infrastructure, such as reliable data networks and power supplies, which may not always be fully supported on-site.

In the Spatial Coordination phase, these barriers and challenges can complicate the integration of robots into the project, requiring careful planning, robust data management, and close collaboration between all project stakeholders to ensure that the benefits of robotics are realized without introducing new risks or inefficiencies.

9.2.2.5 *Stage 4: Technical Design*

High Level of Detail Required: Robots require precise and detailed technical specifications to operate effectively, which can be challenging to provide during the technical design phase, especially when dealing with complex or bespoke design elements.

Complexity in Programming and Customization: The programming and customization required to enable robots to execute specific tasks can be complex and time-consuming, potentially leading to delays in the technical design process.

Interdisciplinary Coordination Issues: Ensuring that robots can work seamlessly with designs from multiple disciplines (e.g., architecture, structural, MEP) requires close coordination, which can be challenging to achieve and maintain throughout the technical design phase.

Limited Flexibility for Design Changes: Once technical designs are finalized for robotic implementation, making changes can be difficult and costly, as it may require reprogramming the robots and adjusting the design specifications.

High Initial Costs: The costs associated with integrating robots into the technical design phase, including software, hardware, and specialized personnel, can be significant, which may strain the project budget.

Training and Expertise Gaps: Technical design teams may lack the necessary expertise to effectively integrate and optimize robotic systems, leading to potential inefficiencies or reliance on external specialists, which can add to project costs and complexity.

Interoperability Challenges: Ensuring that robots can interpret and work with data from various design software tools can be difficult, particularly if there are interoperability issues between the robot's programming and the design outputs.

Complexity in Ensuring Accuracy: The precision required in the technical design phase is critical for robotic operations. Any inaccuracies or discrepancies in the design data can lead to errors in robotic execution, resulting in costly rework.

Potential for Increased Design Time: The need to design with robotics in mind, including the creation of robot-friendly components and systems, can increase the time required to complete the technical design, potentially impacting project timelines.

Reliance on Advanced Infrastructure: The effective use of robots in technical design relies on advanced digital infrastructure, including high-performance computing and reliable data networks, which may not always be readily available or fully integrated into existing workflows.

In the Technical Design phase, these barriers and challenges can complicate the integration of robots, requiring careful planning, investment in training and infrastructure, and close coordination between all project stakeholders to ensure that the use of robotics enhances rather than hinders the design process.

9.2.2.6 *Stage 5: Manufacturing and Construction*

High Initial Investment Costs: Implementing robots in manufacturing and construction involves significant upfront costs for purchasing, programming, and maintaining the robotic systems, which can strain project budgets.

Integration with Existing Processes: Integrating robots into traditional construction workflows can be challenging, as it may require rethinking and restructuring established processes, leading to potential disruptions and inefficiencies.

Complexity in Programming and Operation: Robots often require complex programming and precise operation, which can be time-consuming and necessitate specialized skills that may not be readily available on-site.

Limited Flexibility: Robots are typically designed for specific tasks, making them less adaptable to changes or unexpected conditions on the construction site, which can limit their effectiveness in dynamic environments.

Site Conditions and Environmental Constraints: Construction sites can be unpredictable and varied, with uneven terrain, weather conditions, and other environmental factors that can pose challenges for robots, which are often better suited to controlled environments.

Safety Concerns: Ensuring the safe operation of robots alongside human workers is a significant challenge, requiring careful planning and the implementation of robust safety protocols to prevent accidents or injuries.

Dependence on Digital Infrastructure: The successful deployment of robots relies on a robust digital infrastructure, including reliable internet connectivity and power supplies, which may not always be available or consistent on construction sites.

Training and Skill Gaps: Construction workers may require additional training to operate and maintain robots effectively, which can be costly and time-consuming, and may face resistance from workers unfamiliar with the technology.

Interoperability Issues: Robots may face challenges in integrating with other technologies used on-site, such as BIM or construction management software, leading to potential inefficiencies and data silos.

Potential for Job Displacement: The introduction of robots in construction can lead to concerns about job displacement among workers, potentially causing resistance to the adoption of robotic technology and impacting workforce morale.

Maintenance and Downtime: Robots require regular maintenance and are susceptible to breakdowns, which can lead to downtime and delays in the construction schedule if not managed effectively.

In the Manufacturing and Construction phase, these barriers and challenges can limit the effectiveness of robots, requiring careful planning, investment in training, and the development of robust infrastructure and processes to ensure that the benefits of robotics are fully realized.

9.2.2.7 Stage 6: Handover

Complexity in Data Transfer: Robots used during construction generate vast amounts of data. Ensuring that these data are accurately transferred to facility management systems during the handover can be challenging, potentially leading to data loss or misinterpretation.

Limited Compatibility with Facility Management Systems: The data and models produced by robots may not be fully compatible with the facility management systems in place, leading to difficulties in integrating robotic outputs into the ongoing building management process.

High Costs of Customization: Customizing robots to gather specific as-built data for handover purposes can be expensive, particularly if the robot needs to be reprogrammed or adapted for unique project requirements.

Training Requirements for Facility Managers: Facility managers may need specialized training to understand and use the data collected by robots effectively, which can be a barrier to efficient handover if this training is not provided or if the technology is too complex.

Ensuring Data Accuracy and Completeness: Verifying that the data collected by robots are complete and accurate before handover can be difficult, especially if there were changes during construction that were not fully captured by the robotic systems.

Potential for Misalignment of Expectations: Clients and facility managers may have expectations about the level of detail and accuracy provided by robots, and if these expectations are not met, it can lead to dissatisfaction and potential disputes during handover.

Reliance on Advanced Technology: The use of robots during the handover phase requires reliable technology and infrastructure. Any technical issues or malfunctions can delay the handover process and impact the quality of the delivered data.

Integration Challenges with Existing Workflows: Incorporating robotic data into traditional handover workflows can be complex, especially if the handover process is not fully digitized, leading to potential inefficiencies or duplicated efforts.

Data Overload: The detailed information collected by robots might overwhelm facility managers if not properly organized or filtered, making it difficult to focus on the most critical data for ongoing building operations.

Long-Term Maintenance of Robotic Systems: If robots are intended to be used post-handover for ongoing maintenance tasks, the long-term maintenance and support of these systems can be a challenge, especially if the original construction team is no longer involved.

In the Handover phase, these barriers and challenges can complicate the effective use of robots, requiring careful planning, clear communication, and robust support systems to ensure a smooth and successful transition to building operation and management.

9.2.2.8 Stage 7: Use

High Maintenance and Operational Costs: The ongoing costs associated with maintaining and operating robots during the building's use phase can be substantial, particularly if the robots require specialized parts, regular software updates, or expert technicians.

Integration with Existing Building Management Systems: Robots may face challenges integrating with existing building management systems, leading to potential inefficiencies or the need for costly upgrades to ensure seamless operation.

Complexity in Training Staff: Facility management and maintenance staff may require specialized training to effectively operate and maintain robots, which can be time-consuming and costly, and may face resistance from staff unfamiliar with the technology.

Dependence on Robust Infrastructure: The effective use of robots during the building's use phase depends on a reliable infrastructure, including consistent power supply, internet connectivity, and data networks, which may not always be available or stable.

Potential for Technological Obsolescence: Rapid advancements in robotics technology can lead to obsolescence, where robots used in the building may quickly become outdated, necessitating additional investments in upgrades or replacements.

Safety and Security Concerns: Ensuring the safe operation of robots in an occupied building is critical, particularly in environments with high foot traffic or vulnerable occupants, requiring robust safety protocols and potentially limiting where and how robots can be used.

Data Privacy Issues: Robots equipped with sensors and cameras may collect data during the building's use phase, raising concerns about data privacy and security, particularly in sensitive environments such as healthcare or residential buildings.

Difficulty in Adapting to Changing Needs: Buildings often evolve in use, with spaces being repurposed or reconfigured. Robots may struggle to adapt to these changes, requiring reprogramming or physical modifications to continue operating effectively.

Reliance on Manufacturer Support: The continued use of robots may require ongoing support from the manufacturer, leading to dependency on external vendors for updates, repairs, and technical assistance, which can be costly and lead to delays if issues arise.

Resistance to Adoption: Occupants and facility managers may resist the continued use of robots due to concerns about job displacement, lack of familiarity, or perceived complexity, potentially limiting the effectiveness of robotic systems in the building.

In the Use phase, these barriers and challenges can limit the effectiveness and adoption of robots, requiring careful planning, ongoing support, and proactive management to ensure that the benefits of robotics are fully realized throughout the building's operational life.

9.2.3 *Barriers to Adoption of Drones*

9.2.3.1 *Stage 0: Strategic Definition*

Public Perception as a Barrier: Negative public and stakeholder perception regarding drone use – fuelled by concerns over privacy, safety, and noise – can impede the strategic decision to adopt drone technology in projects.

Regulatory Complexities: The intricate and varying airspace regulations across different jurisdictions can pose a significant barrier, making it difficult to incorporate drone technology into the project's strategic planning.

Regulatory Compliance: Drones are subject to strict regulations regarding their operation, particularly in urban areas or near sensitive sites, which can limit their use during the strategic definition phase and add complexity to project planning.

Privacy Concerns: The use of drones may raise privacy concerns, particularly in densely populated areas, leading to potential resistance from stakeholders or the public and complicating their deployment in the early planning stages.

High Initial Costs: Investing in drones and the necessary training or hiring specialized operators can be costly, which may be difficult to justify during the strategic definition phase when project budgets are still being established.

Limited Data Interpretation Skills: The data collected by drones, such as aerial imagery or topographical surveys, may require specialized skills to interpret effectively, potentially limiting its usefulness in the strategic phase if such expertise is not readily available.

Weather and Environmental Constraints: Drones are highly sensitive to weather conditions, such as strong winds or heavy rain, which can limit their ability to operate reliably during the strategic definition phase and potentially delay data collection.

Integration with Existing Planning Tools: Integrating drone data with traditional planning tools and processes can be challenging, particularly if there are compatibility issues or if the project team lacks experience with drone technology.

Scope of Data Collection: During the strategic definition phase, the scope of data needed may not be fully clear, leading to challenges in defining the specific role of drones and ensuring that the data collected is relevant and useful.

Public Perception and Acceptance: The use of drones can generate negative public perception, especially if there are concerns about surveillance or noise, which could lead to resistance from local communities or stakeholders during the strategic phase.

Security Risks: Drones can be vulnerable to hacking or interference, which poses security risks, particularly if they are being used to collect sensitive data or operate in restricted areas.

Technology Limitations: The technology of drones may not be advanced enough to meet all the strategic needs of a project, such as providing detailed interior surveys or capturing data in complex environments, limiting their effectiveness at this stage.

In the Strategic Definition phase, these barriers and challenges can limit the effective use of drones, requiring careful consideration of their potential impact, integration, and acceptance to ensure they contribute positively to the project's early planning and decision-making processes.

9.2.3.2 Stage 1: Preparation and Brief

Privacy and Security Concerns: The potential for drones to infringe on individual privacy and cause security risks needs to be addressed early. This barrier can complicate the project brief and necessitate the inclusion of strict operational guidelines.

Ethical and Environmental Considerations: Ethical challenges, such as worker surveillance and environmental issues like noise pollution, can limit the acceptance of drones in the early planning stages, hindering the integration of this technology into the project.

Regulatory Restrictions: Drones are subject to regulations that may limit their use during the preparation and brief phase, especially in urban areas, near airports, or in restricted zones, complicating their deployment for site surveys or data collection.

Privacy and Ethical Concerns: The use of drones can raise privacy and ethical concerns, particularly when operating over populated areas or private property, leading to potential resistance from stakeholders and communities.

Data Integration Challenges: Integrating drone-collected data with traditional planning and design tools can be challenging, especially if the project team lacks the technical expertise to process and interpret aerial imagery or 3D models effectively.

Weather-Dependent Operations: Drones are highly dependent on weather conditions for safe and effective operation. Adverse weather such as rain, wind, or fog can delay data collection, potentially impacting project timelines.

High Initial Costs: The costs associated with purchasing, operating, and maintaining drones, as well as training personnel, can be significant, making it difficult to justify their use during the early stages of project development when budgets are tight.

Limited Battery Life and Range: Drones have limited battery life and operational range, which can restrict their ability to cover large sites or conduct prolonged surveys, requiring multiple flights and increasing operational complexity.

Skill and Expertise Requirements: Effective use of drones requires specialized skills and training, which may not be readily available within the project team, necessitating the hiring of external experts and adding to project costs.

Unclear Value Proposition: During the preparation and brief phase, the value that drones add to the project may not be immediately clear, leading to hesitancy in their adoption if stakeholders are unsure of the benefits relative to traditional methods.

Data Overload: Drones can collect large volumes of data, which can be overwhelming to process and analyse, particularly if the project team lacks the tools or expertise to filter and focus on the most relevant information.

Public Perception and Acceptance: The use of drones can generate negative perceptions among the public or stakeholders, particularly if there are concerns about surveillance, noise, or safety, which could lead to opposition or restrictions on their use.

In the Preparation and Brief phase, these barriers and challenges can complicate the effective use of drones, requiring careful planning, stakeholder engagement, and resource allocation to ensure that drones contribute positively to the project's early development and planning stages

9.2.3.3 Stage 2: Concept Design

Limited Use for Detailed Design: Drones are primarily used for capturing external site data and aerial imagery, which may not provide the detailed, precise information needed for the in-depth exploration of design concepts, especially for interior spaces.

Integration Challenges with Design Tools: The data collected by drones, such as aerial photographs or 3D models, may be challenging to integrate with the software tools used for concept design, potentially leading to inefficiencies or requiring additional processing.

Potential Disruption to Creative Process: The reliance on drone data during concept design could limit the creative freedom of designers, as they might feel constrained by the precise site data instead of exploring more innovative or abstract design ideas.

Regulatory and Privacy Constraints: Using drones to capture site data during the concept design phase may be restricted by local regulations or privacy concerns, particularly in urban or sensitive areas, limiting the ability to gather necessary information.

Dependence on External Factors: Drone operations are highly dependent on weather conditions and other external factors, which can delay data collection and, consequently, the progress of the concept design phase.

High Costs for Limited Benefits: The costs associated with using drones for concept design, including equipment, operation, and data processing, may be difficult to justify if the benefits do not significantly enhance the design process or decision-making.

Skill Requirements for Data Interpretation: Interpreting the data collected by drones requires specialized skills, which may not be readily available within the design team, leading to potential delays or the need for external expertise.

Potential for Data Overload: Drones can collect large amounts of data, which may be overwhelming for design teams to process and analyse, particularly when trying to focus on broad design concepts rather than detailed analysis.

Limited Insight into Human Experience: While drones provide valuable site data, they may not effectively capture the human experience or scale, which are crucial aspects of concept design, leading to potential gaps in understanding how users will interact with the space.

Public and Stakeholder Perception: The use of drones during the concept design phase could raise concerns among stakeholders or the public, especially if there are fears about surveillance, noise, or the impact on the surrounding environment.

In the Concept Design phase, these barriers and challenges can limit the effectiveness of drones, requiring careful consideration of their role, integration with design processes, and alignment with the creative goals of the project to ensure they add value without stifling innovation.

Data Integration Challenges: The difficulty in integrating drone-collected data into existing systems and workflows can present a barrier during the concept design phase, requiring advanced data processing capabilities and potentially altering design approaches.

9.2.3.4 Stage 3: Spatial Coordination

Limited Precision for Detailed Coordination: Drones are excellent for capturing general site data but may lack the precision needed for detailed spatial coordination, where millimetre accuracy is often required to avoid clashes between building systems.

Difficulty in Capturing Interior Spaces: Drones are primarily designed for external aerial surveys and may struggle to effectively capture data inside buildings or confined spaces, limiting their usefulness in coordinating interior elements.

Integration with BIM and CAD Tools: The data collected by drones, such as 3D point clouds or aerial imagery, may require significant processing to be integrated with BIM or CAD tools used for spatial coordination, leading to potential delays or the need for specialized software.

Complexity in Managing Large Data Sets: Drones can generate vast amounts of data, which can be challenging to manage and process during the spatial coordination phase, particularly if the project team lacks the necessary tools or expertise.

Reliability of Data Under Environmental Conditions: Environmental factors such as lighting, weather, or site obstructions can affect the quality and reliability of drone data, leading to potential inaccuracies in spatial coordination efforts.

Dependence on Skilled Operators: Effective use of drones for spatial coordination requires skilled operators and data analysts, which may not be readily available, leading to potential delays or the need for additional training.

Regulatory and Safety Constraints: The use of drones in certain areas may be restricted by regulations or safety concerns, limiting their ability to capture the necessary data for spatial coordination, particularly in urban or densely populated areas.

High Costs for Marginal Gains: The investment in drones for spatial coordination might not always yield significant benefits compared to traditional methods, making it difficult to justify the costs associated with their use.

Challenges in Real-Time Data Utilization: Spatial coordination often requires real-time or near-real-time data integration, which can be difficult to achieve with drones due to the time needed for data processing and interpretation.

Limited Applicability for Complex Structures: Drones may struggle to navigate and capture data accurately in highly complex or congested environments, such as dense urban sites or intricate building structures, reducing their effectiveness in coordinating spatial elements.

In the Spatial Coordination phase, these barriers and challenges can limit the utility of drones, necessitating careful planning, appropriate technology integration, and consideration of alternative methods to ensure that spatial coordination is accurate and efficient.

Sensor Interference: Electromagnetic interference affecting drones' gyroscopes and magnetometers can disrupt the accuracy of spatial data collection, posing a significant barrier to their effective use in spatial coordination, particularly in areas near power lines or wireless networks.

Weather Sensitivity: Extreme temperatures and poor weather conditions, such as rain or high winds, can render drones unusable or ineffective, introducing a barrier to site monitoring and surveying during spatial coordination.

9.2.3.5 Stage 4: Technical Design

Skill Gap: A lack of skilled operators proficient in drone technology and data analysis presents a major barrier during the technical design phase, necessitating investment in workforce training and development.

Limited Precision for Detailed Technical Work: Drones may not provide the level of precision required for detailed technical design, where exact measurements and specifications are critical, leading to potential inaccuracies in the design output.

Challenges in Capturing Complex Geometries: Drones may struggle to accurately capture complex geometries or intricate details, which are often required in technical design, limiting their effectiveness in providing the necessary data for these tasks.

Integration with Technical Design Software: The data collected by drones, such as aerial imagery or 3D scans, may not easily integrate with the advanced software used in technical design, necessitating additional processing and potentially causing delays.

Limited Use for Interior Details: Drones are primarily used for exterior data collection and may be less effective at capturing detailed interior information, which is crucial for comprehensive technical design.

Regulatory and Operational Constraints: The use of drones is subject to regulations that may limit their ability to operate in certain areas or at certain times, potentially delaying data collection and impacting the technical design timeline.

High Costs for specialized Data Collection: Using drones for specialized tasks in technical design, such as thermal imaging or high-resolution 3D modelling, can be costly, which might be difficult to justify if the benefits are marginal.

Data Processing and Interpretation Challenges: The raw data collected by drones often require significant processing and interpretation to be useful in technical design, which can be time-consuming and require specialized expertise.

Dependence on Environmental Conditions: The accuracy and quality of drone data can be affected by environmental conditions such as weather, lighting, or obstructions, leading to potential inaccuracies that can compromise the technical design process.

Limited Utility for Small-Scale Projects: For smaller projects or sites, the use of drones may not provide significant advantages over traditional surveying methods, making them less cost-effective and practical in such scenarios.

Reliability and Consistency of Data: Ensuring the reliability and consistency of data collected by drones can be challenging, particularly when multiple flights or different drones are used, leading to potential discrepancies in the technical design phase.

In the Technical Design phase, these barriers and challenges can limit the effectiveness of drones, requiring careful consideration of their capabilities, integration with existing tools, and alignment with the precise requirements of the technical design process to ensure their use adds value without introducing new risks.

9.2.3.6 Stage 5: Manufacturing and Construction

Safety Risks: The safe operation of drones in complex and dynamic construction environments can be challenging, presenting a barrier that requires comprehensive safety protocols and ongoing risk management to avoid accidents.

Weather-Related Operational Delays: Poor weather conditions can severely limit drone operations on construction sites, leading to delays in surveying and monitoring tasks, thus becoming a barrier during the construction phase.

Regulatory Compliance: Drones are subject to strict regulations, especially in urban or restricted areas, which can limit their deployment on construction sites and require additional permits, potentially causing delays.

Weather Dependency: Drones are highly dependent on weather conditions, such as wind, rain, or fog, which can affect their ability to operate reliably and consistently during the construction phase, leading to potential interruptions.

Safety Concerns: Operating drones on active construction sites poses safety risks, particularly in crowded or complex environments where there is a higher chance of collisions with equipment, structures, or workers.

Integration with Construction Workflows: Incorporating drone data into existing construction workflows can be challenging, particularly if the construction team lacks the necessary expertise or tools to effectively use the data collected by drones.

Limited Battery Life and Range: Drones typically have limited battery life and operational range, which can restrict their ability to cover large construction sites or perform prolonged monitoring tasks without frequent recharging or battery changes.

High Initial and Operational Costs: The costs associated with purchasing, maintaining, and operating drones, as well as training personnel, can be significant, making it challenging to justify their use if the perceived benefits do not outweigh these expenses.

Data Processing and Analysis Challenges: The data collected by drones, such as high-resolution images or 3D models, often require extensive processing and analysis, which can be time-consuming and may require specialized software and expertise.

Limited Ability to Perform Detailed Inspections: While drones can provide general site overviews, they may lack the precision needed for detailed inspections of construction elements, particularly in hard-to-reach areas or for close-up assessments.

Dependence on Skilled Operators: Effective use of drones in construction requires skilled operators who are trained in both drone piloting and data interpretation, which may not be readily available on all construction sites, leading to potential operational challenges.

Data Security and Privacy Issues: The use of drones to capture detailed site data can raise concerns about data security and privacy, particularly if the construction site is near sensitive areas or if the data could be accessed by unauthorized individuals.

Potential Resistance from Construction Workers: The introduction of drones on construction sites may face resistance from workers who are concerned about job displacement, privacy, or unfamiliarity with the technology, potentially leading to challenges in adoption and integration.

In the Manufacturing and Construction phase, these barriers and challenges can limit the effectiveness and adoption of drones, requiring careful planning, adherence to regulations, and investment in training and technology to ensure that drones contribute positively to the construction process without introducing new risks or inefficiencies.

9.2.3.7 Stage 6: Handover

Data Integration with Facility Management Systems: The data collected by drones, such as aerial imagery or 3D models, may not easily integrate with existing facility management systems, making it challenging to incorporate these data into the handover package effectively.

Accuracy of As-Built Documentation: While drones can capture external site conditions, they may not provide the level of detail needed for accurate as-built documentation, particularly for interior spaces or complex structural details, potentially leading to incomplete records.

Regulatory and Compliance Issues: Using drones during the handover phase may raise regulatory or compliance concerns, especially if the site is in a sensitive area or if there are restrictions on drone flights, which could delay the handover process.

Data Security and Privacy Concerns: The use of drones to capture detailed site data can raise security and privacy concerns, especially if the data include sensitive areas of the building or surrounding properties, leading to potential legal challenges.

Dependence on Skilled Operators: Effective use of drones during handover requires skilled operators who can accurately capture and process the necessary data, which may not be readily available, leading to potential delays or additional costs.

Cost Justification: The costs associated with deploying drones during the handover phase, including data processing and analysis, may be difficult to justify if the perceived benefits do not significantly enhance the handover process.

Limited Use for Interior Inspections: Drones are primarily effective for external inspections and may struggle to capture detailed interior conditions, limiting their usefulness in providing a comprehensive assessment of the building during handover.

Potential for Incomplete Data Capture: The reliance on drones for data capture may result in incomplete documentation if certain areas are inaccessible or if the drone's sensors are not capable of capturing specific details needed for the handover.

Challenges in Ensuring Data Consistency: Ensuring that the data captured by drones are consistent and aligns with other documentation can be challenging, particularly if multiple flights or different drones are used, leading to potential discrepancies.

Client Familiarity and Acceptance: Clients may be unfamiliar with drone technology and may question the accuracy or relevance of the data collected by drones, potentially leading to resistance or a lack of confidence in the handover documentation.

In the Handover phase, these barriers and challenges can limit the effectiveness of drones, requiring careful planning, skilled operation, and clear communication to ensure that drones add value to the handover process without compromising data quality or regulatory compliance.

Data Inaccuracy: Any interference with drone sensors during data collection or operational inefficiencies caused by adverse weather can result in inaccurate data, which may serve as a barrier to a smooth and precise handover process.

9.2.3.8 Stage 7: Use

Ethical and Environmental Barriers: During the post-construction phase, continued drone usage can raise ethical issues such as worker surveillance and environmental concerns, creating barriers to their widespread use in building maintenance and monitoring.

Psychological Resistance to Technology: Psychological resistance from workers or stakeholders uncomfortable with the use of drones can persist, acting as a barrier to their full utilization during the operational life of the building.

Ongoing Operational Costs: The continued use of drones during the building's operational phase can incur significant costs, including maintenance, software updates, and the need for skilled operators, which may strain facility management budgets.

Data Privacy and Security Concerns: Drones used during the building's operational phase may raise privacy and security concerns, particularly if they are capturing data in sensitive areas or around occupied spaces, leading to potential legal and compliance issues.

Integration with Existing Facility Management Systems: The data captured by drones may not seamlessly integrate with existing facility management systems, requiring additional effort to process and make use of the information, potentially leading to inefficiencies.

Limited Utility for Indoor Inspections: Drones are generally more effective for outdoor monitoring and inspections. Their use indoors is often limited by space constraints, navigation difficulties, and the inability to capture detailed interior conditions.

Dependence on Environmental Conditions: The effectiveness of drones can be compromised by environmental factors such as weather, lighting, and obstructions, which may limit their utility for regular monitoring or inspections during the building's use phase.

Skill and Training Requirements: Continued use of drones requires ongoing training for facility management staff or reliance on external operators, which can add complexity and cost to building operations.

Potential for Technological Obsolescence: The rapid pace of technological advancement means that drone technology used during the building's use phase may quickly become outdated, necessitating further investment in new equipment or upgrades.

Regulatory Compliance Challenges: Ongoing drone operations must comply with local and national regulations, which can change over time, adding complexity to the process and potentially restricting the use of drones in certain areas or for certain tasks.

Safety Risks in Occupied Spaces: Operating drones in or around occupied buildings poses safety risks, particularly in crowded or public areas, requiring strict safety protocols to prevent accidents or injuries.

Limited Value for Routine Tasks: For routine facility management tasks, the use of drones may not provide significant advantages over traditional methods, making it difficult to justify their continued use during the building's operational phase.

In the Use phase, these barriers and challenges can limit the practical benefits of drones, requiring careful consideration of their role, cost-effectiveness, and integration with existing building management practices to ensure they contribute positively to the ongoing operation and maintenance of the facility.

9.2.4 *Barriers to Adoption of AR/VR/MR*

9.2.4.1 Stage 0: Strategic Definition

High Initial Costs: Implementing AR, VR, and MR technologies requires significant investment in hardware, software, and training, which can be difficult to justify during the strategic definition phase when budgets are still being established.

Limited Stakeholder Familiarity: Many stakeholders, including clients and decision-makers, may be unfamiliar with AR, VR, and MR technologies, leading to resistance or hesitation in adopting these tools during the strategic phase.

Complexity in Demonstrating Value: The strategic phase focuses on high-level goals and objectives, and it can be challenging to clearly demonstrate the tangible value of AR, VR, and MR in achieving these outcomes at such an early stage.

Technology Integration Challenges: Integrating AR, VR, and MR with existing strategic planning tools and processes can be complex, requiring additional effort and expertise that may not be readily available within the project team.

High Learning Curve: The use of AR, VR, and MR technologies often involves a steep learning curve, which can slow down the strategic planning process if key stakeholders or team members need to acquire new skills.

Limited Content Availability: At the strategic definition stage, there may be limited content or data available to create meaningful AR, VR, or MR experiences, reducing the effectiveness of these technologies in informing strategic decisions.

Uncertain ROI: The ROI for AR, VR, and MR technologies may be unclear during the strategic phase, leading to hesitancy in committing resources to these tools without a clear understanding of their long-term benefits.

Potential for Overcomplication: Introducing AR, VR, and MR technologies in the strategic phase can overcomplicate the planning process, diverting attention from core objectives and strategic decision-making.

Technical Limitations: The current capabilities of AR, VR, and MR technologies may not fully align with the needs of the strategic phase, particularly if the focus is on broad, conceptual planning rather than detailed visualizations.

Dependence on Digital Infrastructure: Effective use of AR, VR, and MR requires robust digital infrastructure, including high-speed internet and advanced computing hardware, which may not be available in all settings or for all stakeholders involved in the strategic phase.

In the Strategic Definition phase, these barriers and challenges can limit the effectiveness of AR, VR, and MR technologies, requiring careful consideration of their application, cost, and integration to ensure they add value without complicating the strategic planning process.

9.2.4.2 Stage 1: Preparation and Brief

Cultural and Organizational Resistance: During the briefing stage, resistance to adopting new technologies due to ingrained traditional practices can delay decisions about incorporating AR, VR, and MR into the project. Overcoming this requires addressing the mindset and awareness gap early on.

High Implementation Costs: The costs associated with deploying AR, VR, and MR, including hardware, software, and training, can be significant, which may be difficult to justify during the preparation and brief phase when budgets are still being allocated.

Limited Stakeholder Familiarity: Many clients and stakeholders may be unfamiliar with AR, VR, and MR technologies, leading to potential resistance or a lack of engagement with these tools during the early stages of the project.

Challenges in Capturing Client Requirements: Accurately translating client requirements into AR, VR, or MR environments can be complex, potentially leading to miscommunication or misunderstandings if the technology is not used effectively.

Complexity in Customization: Customizing AR, VR, and MR experiences to meet the specific needs of the project can be time-consuming and technically challenging, requiring specialized skills that may not be readily available.

Interoperability Issues: Integrating AR, VR, and MR tools with existing project management and design software can be challenging, potentially leading to data compatibility issues and inefficiencies in the preparation phase.

Steep Learning Curve: The use of AR, VR, and MR technologies often involves a steep learning curve for project teams, which can slow down the preparation and briefing process if key participants are not familiar with the tools.

Data Accuracy and Relevance: Ensuring that the data and models used in AR, VR, and MR environments are accurate and relevant to the project's needs is crucial, but this can be difficult to achieve in the early stages when information is still being gathered.

Overreliance on Technology: There is a risk of overreliance on AR, VR, and MR at the expense of traditional methods, which may lead to a focus on visualizations rather than the core strategic and functional aspects of the project brief.

Accessibility and Inclusivity: Not all stakeholders may have equal access to AR, VR, and MR technologies, potentially leading to inclusivity challenges where certain participants are unable to fully engage with the briefing process.

Risk of Overcomplicating the Briefing Process: The introduction of AR, VR, and MR technologies can add complexity to the briefing process, potentially leading to confusion or delays if the technology is not seamlessly integrated into the workflow.

In the Preparation and Brief phase, these barriers and challenges can limit the effectiveness of AR, VR, and MR technologies, requiring careful planning, stakeholder education, and strategic integration to ensure they add value without complicating the process.

9.2.4.3 Stage 2: Concept Design

High Cost of Implementation: The costs associated with AR, VR, and MR technology, including hardware, software, and skilled personnel, can be substantial, making it challenging to justify their use during the concept design phase when budgets are still being defined.

Technical Complexity: Creating and managing AR, VR, and MR environments can be technically complex, requiring specialized skills that may not be readily available within the design team, potentially leading to delays or increased project costs.

Steep Learning Curve: Designers and stakeholders may face a steep learning curve when adopting AR, VR, and MR tools, which can slow down the design process and reduce the efficiency of design exploration.

Potential for Overemphasis on Visuals: The immersive and visually appealing nature of AR, VR, and MR can lead to an overemphasis on aesthetics at the expense of functionality, feasibility, and other critical design considerations.

Integration with Traditional Design Processes: Integrating AR, VR, and MR with traditional design workflows can be challenging, particularly if the tools do not easily interface with existing design software, leading to potential inefficiencies or data loss.

Data Accuracy and Reliability: Ensuring that the data and models used in AR, VR, and MR environments are accurate and up-to-date is crucial, but maintaining this accuracy can be difficult during the rapidly evolving concept design phase.

Potential for Design Fixation: The immersive nature of AR, VR, and MR can lead to design fixation, where designers become overly attached to a particular concept or visual representation, limiting creativity and exploration of alternative solutions.

Limited Stakeholder Accessibility: Not all stakeholders may have access to the necessary AR, VR, or MR technology, which can create barriers to participation and limit the inclusivity of the design process.

Overwhelming Complexity for Clients: The advanced visualizations provided by AR, VR, and MR can overwhelm clients, particularly if they are not familiar with these technologies, leading to confusion or miscommunication about the design intent.

Risk of Technology Dependency: Overreliance on AR, VR, and MR technologies can lead to a reduction in traditional design skills and methods, potentially weakening the overall robustness and versatility of the design process.

Technological Challenges (Integration with BIM): At this stage, design concepts are being developed, and integration with digital tools like BIM is critical. Issues related to data accuracy and technology integration with AR, VR, and MR systems may pose barriers.

In the Concept Design phase, these barriers and challenges can limit the effectiveness of AR, VR, and MR technologies, requiring careful consideration of their application, strategic integration with traditional methods, and clear communication with all stakeholders to ensure they enhance rather than hinder the design process.

9.2.4.4 Stage 3: Spatial Coordination

Data Integration Challenges: Integrating AR, VR, and MR with BIM and other spatial coordination tools can be complex, requiring seamless data transfer and interoperability between different platforms, which is not always guaranteed.

High Cost of Implementation: The deployment of AR, VR, and MR technologies for spatial coordination can involve significant costs for hardware, software, and skilled personnel, which may be challenging to justify within project budgets.

Steep Learning Curve: Effective use of AR, VR, and MR in spatial coordination requires specialized skills and training, which can slow down the adoption process and reduce the efficiency of project teams not familiar with these technologies.

Potential for Technical Issues: AR, VR, and MR technologies can be prone to technical glitches or software compatibility issues, leading to potential delays in the spatial coordination process and the risk of misalignment in the coordinated models.

Overreliance on Technology: There is a risk that overreliance on AR, VR, and MR could overshadow traditional coordination methods, potentially leading to a lack of critical thinking or manual checks that are essential for ensuring spatial accuracy.

User Accessibility and Inclusivity: Not all team members or stakeholders may have access to AR, VR, or MR technologies, leading to challenges in ensuring that everyone can participate equally in the spatial coordination process.

Complexity in Managing Large Data Sets: The detailed models used in AR, VR, and MR environments can generate large data sets that are difficult to manage, store, and process, potentially leading to inefficiencies in handling the spatial coordination data.

Potential for Visual Misinterpretation: While AR, VR, and MR provide immersive visualizations, there is a risk that users may misinterpret the spatial relationships or scale within these environments, leading to coordination errors.

Time-Consuming Setup and Calibration: Setting up and calibrating AR, VR, and MR systems for spatial coordination can be time-consuming, potentially delaying the coordination process, especially if the technology requires frequent adjustments.

Dependence on High-Quality Input Data: The effectiveness of AR, VR, and MR in spatial coordination is heavily dependent on the accuracy and quality of the input data. Any inaccuracies in the initial data can lead to compounded errors in the coordinated output.

Interoperability Issues: During spatial coordination, effective use of AR, VR, and MR technologies may be hindered by interoperability issues with other digital systems such as BIM and GIS. These limitations affect seamless data exchange and visualization across platforms.

In the Spatial Coordination phase, these barriers and challenges can limit the effectiveness of AR, VR, and MR technologies, requiring careful planning, adequate training, and robust data management to ensure that these tools enhance spatial accuracy and project efficiency without introducing new risks or complications.

9.2.4.5 *Stage 4: Technical Design*

Lack of Expertise and Training: As technical design is developed, the need for trained personnel to effectively utilize AR, VR, and MR technologies becomes more apparent. This barrier reflects the skills gap that must be addressed to ensure successful implementation.

High Implementation Costs: The deployment of AR, VR, and MR technologies during the technical design phase involves substantial costs, including hardware, software, and training, which can strain project budgets and limit their widespread adoption.

Complexity in Creating Detailed Models: Developing detailed and accurate AR, VR, or MR models for technical design can be complex and time-consuming, requiring advanced skills in 3D modelling and data management, which may not be readily available within the project team.

Steep Learning Curve: Designers and engineers may face a steep learning curve when adopting AR, VR, and MR tools, potentially slowing down the design process and affecting productivity if team members are not adequately trained.

Integration with Existing Design Tools: Integrating AR, VR, and MR technologies with traditional CAD and BIM tools can be challenging, especially if there are compatibility issues or if the existing design workflows are not optimized for these technologies.

Risk of Data Overload: The use of AR, VR, and MR can generate vast amounts of detailed data, which can overwhelm designers and complicate the technical design process if not managed effectively.

Potential for Overreliance on Visuals: The immersive and highly visual nature of AR, VR, and MR may lead to an overemphasis on aesthetics, potentially diverting attention from critical technical details, specifications, and functional requirements.

Limited Effectiveness for Complex Technical Details: While AR, VR, and MR are excellent for visualization, they may struggle to convey highly complex technical details, such as precise tolerances, material properties, or engineering calculations, which are crucial in technical design.

Challenges in Collaboration: Collaborative design using AR, VR, and MR can be challenging if all team members do not have access to the necessary technology or if there are issues with synchronizing data in real-time across different platforms.

Dependence on High-Quality Data: The effectiveness of AR, VR, and MR in the technical design phase relies on the accuracy and completeness of the underlying data. Any errors or gaps in the data can lead to inaccuracies in the final design output.

Potential for Technology Dependency: Overreliance on AR, VR, and MR could lead to a decrease in traditional technical design skills, as designers may become dependent on these tools for tasks that were previously done manually, potentially reducing overall design robustness.

In the Technical Design phase, these barriers and challenges can limit the practical benefits of AR, VR, and MR technologies, requiring careful planning, robust data management, and a balanced approach that integrates these tools effectively without compromising the quality and precision of the design work.

9.2.4.6 *Stage 5: Manufacturing and Construction*

Technological Challenges (Hardware): AR, VR, and MR devices need to be rugged enough for construction site conditions. Additionally, a stable WLAN and internet connection are vital for real-time functionality, posing challenges during the manufacturing and construction phases.

High Implementation and Operational Costs: The deployment of AR, VR, and MR technologies in manufacturing and construction involves significant costs, including purchasing specialized hardware, developing custom software, and training personnel, which can strain project budgets.

Technical Integration Challenges: Integrating AR, VR, and MR with existing construction technologies and workflows can be complex, potentially leading to compatibility issues, inefficiencies, and delays if not managed effectively.

Limited On-Site Expertise: Construction teams may lack the necessary expertise to effectively use AR, VR, and MR technologies on-site, requiring additional training or the hiring of specialized personnel, which can disrupt construction timelines.

Dependence on Stable Digital Infrastructure: The effectiveness of AR, VR, and MR on construction sites depends on reliable digital infrastructure, including high-speed internet and power supplies, which may not always be available, especially in remote or challenging environments.

Potential for Technology Disruptions: Technical issues such as software glitches, hardware malfunctions, or connectivity problems can disrupt the use of AR, VR, and MR on-site, leading to potential delays and increased costs.

Health and Safety Concerns: The use of AR, VR, and MR on active construction sites can raise safety concerns, particularly if workers are distracted by or overly reliant on these technologies, potentially leading to accidents or errors.

Complexity in Real-Time Data Integration: Integrating real-time data from AR, VR, and MR systems with on-site construction activities can be challenging, particularly when trying to synchronize updates across multiple devices and teams.

Resistance to Adoption: Construction workers and site managers may resist adopting AR, VR, and MR technologies due to unfamiliarity, concerns about job displacement, or scepticism about the technology's practical benefits, hindering their widespread use.

Difficulty in Visualizing Complex Construction Processes: While AR, VR, and MR are powerful visualization tools, they may struggle to accurately represent complex, dynamic construction processes or to convey the detailed step-by-step instructions needed for certain tasks.

Time-Consuming Setup and Calibration: Setting up and calibrating AR, VR, and MR systems for use on construction sites can be time-consuming, potentially causing delays in the construction schedule if not planned and executed efficiently.

In the Manufacturing and Construction phase, these barriers and challenges can limit the effectiveness and adoption of AR, VR, and MR technologies, requiring careful planning, thorough training, and robust infrastructure to ensure that these tools enhance construction efficiency and safety without introducing new risks or complications.

9.2.4.7 Stage 6: Handover

Legal and Regulatory Uncertainties: At handover, legal issues, including liability concerns, data privacy, and intellectual property rights associated with AR, VR, and MR use, may complicate the process. Clear guidelines and frameworks are required to ensure smooth technology handover.

Complexity in Data Transfer and Integration: AR, VR, and MR technologies generate large amounts of data that may be difficult to integrate into the client's existing facility management systems, potentially leading to gaps in information or inefficiencies in the handover process.

High Costs for Implementation: The costs associated with creating and utilizing AR, VR, and MR tools during handover, including hardware, software, and training, can be significant, potentially making it challenging to justify their use in this phase.

Limited Client Familiarity: Clients may be unfamiliar with AR, VR, and MR technologies, leading to resistance or difficulty in understanding and effectively using these tools for building operations and maintenance post-handover.

Training and Skill Gaps: Facility management teams may require specialized training to effectively use AR, VR, and MR systems provided during handover, which can be time-consuming and costly, and may result in underutilization if not addressed.

Technical Compatibility Issues: Ensuring that the AR, VR, and MR data and systems are compatible with the client's existing digital infrastructure can be challenging, leading to potential delays or the need for additional resources to resolve compatibility issues.

Data Overload: The extensive amount of information provided through AR, VR, and MR can overwhelm clients or facility managers, making it difficult for them to extract and focus on the most critical data needed for effective building operation.

Security and Privacy Concerns: The detailed digital models and data used in AR, VR, and MR applications can raise security and privacy concerns, especially if the handover includes sensitive information about the building's infrastructure.

Potential for Incomplete Documentation: There is a risk that not all aspects of the building will be fully documented in AR, VR, or MR formats, particularly if there were changes or updates made late in the construction process that were not captured.

Dependence on Continuous Technology Support: The client may need ongoing technical support to maintain and update AR, VR, and MR systems, leading to additional costs and potential dependency on external vendors, which can complicate long-term building management.

Unclear Long-Term Value: The long-term benefits of using AR, VR, and MR in the handover phase may be unclear to clients, especially if they do not plan to continue using these technologies for building management, leading to concerns about the ROI.

In the Handover phase, these barriers and challenges can limit the effectiveness of AR, VR, and MR technologies, requiring careful planning, client education, and strategic integration to ensure they add value without complicating the transition to building operation and management.

9.2.4.8 Stage 7: Use

Technological Challenges (Ongoing Maintenance): Long-term use of AR, VR, and MR technologies requires continuous updates and maintenance. Hardware and software limitations that impact real-time updates and performance can become barriers during this operational phase.

High Ongoing Costs: The continued use of AR, VR, and MR technologies requires ongoing investment in hardware maintenance, software updates, and training, which can be expensive and may strain operational budgets.

Limited User Adoption: Building occupants, facility managers, and maintenance staff may be reluctant to adopt AR, VR, and MR technologies, especially if they are unfamiliar with these tools or perceive them as complex or unnecessary.

Technical Skill Requirements: Effective use of AR, VR, and MR in the building's operational phase requires specialized skills, which may not be readily available among the facility management team, necessitating additional training or hiring of experts.

Integration Challenges with Existing Systems: AR, VR, and MR technologies may not easily integrate with existing facility management and building automation systems, leading to potential inefficiencies or the need for costly custom solutions.

Dependence on Robust Infrastructure: The effectiveness of AR, VR, and MR relies on a stable digital infrastructure, including reliable internet connectivity and sufficient computing power, which may not always be available or consistently maintained in the long term.

Data Security and Privacy Concerns: The use of AR, VR, and MR involves the handling of detailed building data, which raises concerns about data security and privacy, especially in sensitive environments such as hospitals, government buildings, or residential properties.

Risk of Technology Obsolescence: Rapid advancements in AR, VR, and MR technologies could render existing systems obsolete, requiring frequent upgrades or replacements to maintain their effectiveness, leading to potential disruptions and additional costs.

Complexity in Managing and Updating Digital Models: Keeping digital models used in AR, VR, and MR applications up to date with ongoing building modifications or changes can be challenging, leading to potential discrepancies between the virtual models and the physical building.

Overreliance on Technology: There is a risk that facility managers may become overly reliant on AR, VR, and MR technologies, potentially neglecting traditional skills and methods that are crucial for effective building management.

Limited Accessibility for All Users: Not all building users may have access to AR, VR, and MR technologies, either due to cost, physical limitations, or lack of familiarity, leading to potential exclusion or reduced effectiveness in operations.

Potential for Distraction or Misuse: The immersive nature of AR, VR, and MR could lead to distractions or misuse, especially in environments where focus and precision are critical, such as in high-security or high-risk areas of the building.

In the Use phase, these barriers and challenges can limit the practical benefits of AR, VR, and MR technologies, requiring careful consideration of their implementation, ongoing support, and integration with existing systems to ensure they enhance building operations without introducing new risks or complexities.

9.2.5 Barriers to the Adoption of 3D Printers

The adoption of 3D printing technology in the construction sector represents a paradigm shift, offering unprecedented opportunities for innovation, efficiency, and sustainability. Despite these potential benefits, the integration of 3D printing into construction projects faces significant barriers. This essay explores these obstacles, including technical challenges, regulatory and standardization issues, economic and financial constraints, workforce and skills gaps, cultural and organizational resistance, supply chain and material limitations, safety and quality concerns, environmental impacts, and the need for research and development (R&D). Understanding and addressing these barriers is crucial for harnessing the full potential of 3D printing in construction.

Technical Challenges: One of the primary barriers to the adoption of 3D printers in construction is the technical limitations of current technology. While 3D printing has seen substantial advancements, scalability issues persist, particularly in printing large structures or buildings within feasible timeframes. Additionally, the precision and reliability of 3D printers need to be improved to meet the rigorous standards of construction projects. For instance, Dörfler et al. (2024) highlight that current 3D printing technologies struggle with the variability in environmental conditions on construction sites, affecting the consistency and integrity of printed materials.

Regulatory and Standardization Issues: The lack of specific regulations and standards for 3D printing in construction poses a significant barrier to its adoption. The construction industry is highly regulated, and new technologies must comply with existing building codes and standards. However, the novel nature of 3D printing means that regulatory frameworks have yet to catch up, creating uncertainty for companies wishing to adopt this technology. As Sovetova and Calautit (2024) note, standardization in terms of material properties, printing processes, and structural performance is essential for the broader acceptance of 3D printing in construction.

Economic and Financial Constraints: The initial investment required for 3D printing technology, including the printers themselves and the necessary infrastructure, is substantial. Furthermore, the cost-effectiveness of 3D printing for construction projects is still under debate, with concerns over the total cost of ownership, including maintenance, materials, and operation costs. A report by McKinsey & Company (2019) suggests that while 3D printing can reduce labour costs and waste, these savings must be balanced against the high initial investment and ongoing costs.

Workforce and Skills Gaps: The integration of 3D printing into construction necessitates a workforce skilled in both traditional construction techniques and new digital fabrication methods. Currently, there is a significant skills gap in the industry, with a shortage of professionals trained in 3D printing technologies. This gap hinders the adoption of 3D printing, as firms struggle to find or train employees capable of operating the new technology efficiently. According to Ghosh and Karmakar (2024), upskilling the existing workforce and incorporating 3D printing into construction education and training programmes are crucial steps toward overcoming this barrier.

Cultural and Organizational Resistance: Cultural and organizational resistance to change is a pervasive barrier in the construction industry, known for its conservatism and reliance on established practices. The introduction of 3D printing challenges traditional construction methodologies and requires a shift in mindset from all stakeholders involved. This resistance to change can slow the adoption of 3D printing, as noted by Ford and Despeisse (2016), who argue that overcoming this barrier requires demonstrating the tangible benefits of 3D printing, such as increased efficiency and innovation, to all stakeholders.

Supply Chain and Material Limitations: The supply chain for materials suitable for 3D printing in construction is currently underdeveloped, with a limited range of materials available that meet the required standards for strength, durability, and environmental performance. Additionally, the logistics of supplying these materials to construction sites in a form ready for 3D printing presents challenges. Research into new materials and the development of a robust supply chain are essential for the broader adoption of 3D printing in construction, as highlighted by Rogers et al. (2024).

Safety and Quality Concerns: Safety and quality assurance are paramount in construction, and any new technology must meet strict criteria in these areas. Concerns over the structural integrity and long-term performance of 3D-printed buildings need to be addressed to gain the confidence of builders, regulators, and end-users. The development of rigorous testing and certification processes for 3D-printed structures is crucial, which emphasizes the need for comprehensive standards covering the design, printing process, and material properties.

Environmental Impacts: While 3D printing has the potential to reduce waste and improve sustainability in construction, concerns about the environmental impact of printing materials and the energy consumption of 3D printers persist. Assessing and mitigating the environmental footprint of 3D printing in construction requires further research into eco-friendly materials and energy-efficient printing technologies.

R&D: The advancement of 3D printing technology for construction applications requires ongoing R&D. However, the construction industry's investment in R&D is historically low compared to other sectors, limiting the resources available for developing and refining 3D printing technologies. Increased investment in R&D is essential for overcoming the technical, material, and process-related barriers to the adoption of 3D printing in construction, as argued by Oesterreich and Teuteberg (2016).

9.2.5.1 Stage 0: Strategic Definition

High Initial Costs: The investment required for 3D printing technology, including purchasing printers, materials, and related software, can be substantial, making it challenging to justify during the early strategic planning stage when budgets are still being defined.

Limited Strategic Value Perception: At the strategic definition phase, the value of 3D printing may not be immediately clear to stakeholders, leading to scepticism or reluctance to incorporate this technology into the project plan.

Technology Complexity: Understanding the capabilities and limitations of 3D printing requires technical expertise, which may not be available during the strategic phase, leading to potential underestimation or overestimation of its benefits.

Integration Challenges: Integrating 3D printing into the broader project strategy can be complex, especially when aligning it with traditional construction methods and processes, which may require adjustments or additional planning.

Uncertain ROI: The ROI for 3D printing may be uncertain during the strategic phase, particularly if the long-term benefits are difficult to quantify or if the technology's potential is not fully understood by decision-makers.

Material and Process Limitations: The materials and processes available for 3D printing may have limitations that are not fully compatible with the strategic goals of the project, such as durability concerns or restrictions on the types of structures that can be printed.

Potential for Misalignment with Project Goals: The introduction of 3D printing at the strategic phase may lead to a misalignment with the overall project goals if its capabilities are not accurately assessed or if it is not integrated into the project's broader objectives.

Regulatory and Compliance Concerns: There may be regulatory or compliance challenges associated with the use of 3D printing in construction, particularly if the technology is ahead of existing building codes or standards, which could delay project approval.

Environmental and Sustainability Considerations: While 3D printing has the potential for sustainability benefits, there may be concerns about the environmental impact of certain materials or the energy consumption of the printing process, which need to be considered during strategic planning.

Dependence on Technological Advances: The strategic use of 3D printing may be hindered by the pace of technological advancements, with the risk that current capabilities may quickly become outdated, requiring ongoing investment in newer technology.

In the Strategic Definition phase, these barriers and challenges can limit the adoption and integration of 3D printing into the project, requiring careful evaluation of its potential, alignment with project goals, and consideration of long-term benefits and risks.

9.2.5.2 *Stage 1: Preparation and Brief*

High Initial Setup Costs: The cost of acquiring and setting up 3D printers, along with necessary materials and software, can be significant, potentially straining the budget during the early preparation and briefing stage.

Limited Expertise: The effective use of 3D printing technology requires specialized skills and knowledge, which may not be available within the project team during the preparation phase, leading to potential delays or reliance on external experts.

Challenges in Communicating Design Intent: While 3D models can be useful, they may not fully capture the design intent or convey the project's strategic goals as effectively as traditional drawings or digital models, leading to potential miscommunication with stakeholders.

Integration with Traditional Methods: Incorporating 3D printing into established workflows can be challenging, especially if the project team is more familiar with traditional design and construction methods, leading to potential inefficiencies or resistance.

Material Limitations: The range of materials that can be used in 3D printing is still limited, which may restrict the ability to create models that accurately represent the final construction materials and their properties.

Scale and Detail Limitations: 3D printers may struggle to produce large-scale models or highly detailed components, which can limit their usefulness in the preparation and brief phase where comprehensive and precise representations are needed.

Time-Consuming Process: The process of 3D printing, especially for complex or large models, can be time-consuming, potentially delaying the project timeline during a phase when speed and efficiency are critical.

Potential for Overreliance: There is a risk of overreliance on 3D printing technology, which could lead to the neglect of other important aspects of the preparation and brief phase, such as stakeholder engagement or detailed planning.

Difficulty in Iterative Design: Iterating on designs using 3D printing can be slower and more resource-intensive compared to digital modelling, which may hinder the flexibility needed during the preparation and briefing stage.

Uncertain ROI: The ROI for using 3D printing in the preparation and brief phase may be unclear, particularly if the technology does not significantly enhance the decision-making process or lead to tangible project benefits.

In the Preparation and Brief phase, these barriers and challenges can limit the effectiveness and practicality of 3D printing, requiring careful consideration of when and how to use this technology to ensure it supports, rather than hinders, the project's early development.

9.2.5.3 *Stage 2: Concept Design*

Limited Creative Flexibility: 3D printers may constrain the creative exploration of design concepts due to limitations in the range of shapes, sizes, and materials they can produce, potentially stifling innovation during the concept design phase.

High Costs for Iterative Models: The cost of producing multiple 3D-printed models for iterative design can quickly add up, making it expensive to explore different design options and refinements during the concept stage.

Time-Consuming Production: The process of 3D printing, especially for complex or large-scale models, can be time-consuming, potentially delaying the rapid prototyping and design iteration needed in the concept phase.

Material and Detail Limitations: 3D printers are limited in the types of materials they can use and may struggle to accurately reproduce fine details or textures, which can result in models that do not fully capture the intended design nuances.

Potential for Overemphasis on Physical Models: Relying heavily on 3D-printed models may lead to an overemphasis on physical representations, potentially diverting focus from other important aspects of concept design, such as digital visualization or environmental considerations.

Integration with Digital Design Tools: Integrating 3D-printed models with digital design workflows can be challenging, particularly if the models need to be frequently updated or if there are discrepancies between the digital and physical representations.

Skill and Expertise Requirements: Effective use of 3D printing in concept design requires specialized knowledge and skills, which may not be readily available within the design team, leading to potential delays or reliance on external experts.

Environmental Impact: The production of 3D-printed models can generate waste and consume energy, raising environmental concerns, especially if multiple iterations are required during the concept design phase.

Risk of Design Fixation: The tangible nature of 3D-printed models may lead to design fixation, where designers and stakeholders become overly attached to a specific concept, limiting the exploration of alternative ideas and approaches.

Uncertain Value in Early Stages: The added value of 3D printing in the early stages of design may be unclear, as the conceptual nature of this phase may not require the level of detail or precision that 3D printing offers, making its use potentially redundant.

In the Concept Design phase, these barriers and challenges can limit the practical benefits of 3D printing, requiring careful consideration of its role in the design process to ensure it enhances creativity and innovation without imposing unnecessary constraints or costs.

9.2.5.4 Stage 3: Spatial Coordination

Limited Precision for Detailed Coordination: 3D printers may not provide the level of precision required for spatial coordination, particularly when dealing with complex systems like MEP (mechanical, electrical, and plumbing), where millimetre accuracy is critical.

Challenges in Representing Complex Systems: The ability of 3D printers to accurately represent intricate and densely packed systems can be limited, making it difficult to visualize and coordinate the spatial relationships between different building components.

High Costs for Large-Scale Models: Producing large-scale models that are necessary for effective spatial coordination can be expensive, especially when multiple iterations are needed to refine the coordination of different systems.

Time-Intensive Production: The process of 3D printing detailed and accurate models for spatial coordination can be time-consuming, potentially delaying project timelines if the printing and assembly of models take longer than anticipated.

Material Constraints: The materials used in 3D printing may not fully replicate the properties of real construction materials, which can lead to models that are less effective for spatial coordination, particularly when considering aspects like load-bearing or thermal performance.

Difficulty in Updating Models: Making changes to 3D-printed models in response to design adjustments or coordination issues can be cumbersome, requiring new prints for each iteration, which is less flexible compared to digital models that can be easily modified.

Risk of Overreliance on Physical Models: Relying heavily on 3D-printed models for spatial coordination may limit the use of digital tools that offer more dynamic and interactive ways to explore and resolve coordination issues, such as clash detection in BIM software.

Storage and Handling Challenges: Large or numerous 3D-printed models can be difficult to store and transport, especially in the context of ongoing spatial coordination where models may need to be frequently accessed and reviewed by different teams.

Skill and Expertise Requirements: Effective use of 3D printing in spatial coordination requires specialized knowledge, which may not be readily available, leading to potential delays or the need to involve external experts.

Environmental Impact: The production of 3D-printed models can generate waste and consume significant energy, raising sustainability concerns, particularly when multiple iterations are needed during the spatial coordination phase.

In the Spatial Coordination phase, these barriers and challenges can limit the effectiveness of 3D printing, requiring careful consideration of when and how to use this technology to ensure it enhances coordination efforts without introducing additional costs, delays, or inefficiencies.

9.2.5.5 Stage 4: Technical Design

Limited Precision for Technical Detailing: 3D printers may not achieve the exacting levels of precision required for technical design, where small inaccuracies can lead to significant issues in the final construction, particularly for complex or finely detailed components.

Material Limitations: The range of materials available for 3D printing may not accurately represent the materials specified in the technical design, potentially leading to misinterpretations of how certain elements will perform or appear in the finished building.

High Costs for Detailed Models: Producing detailed and accurate 3D-printed models for technical design can be expensive, especially when multiple iterations are required to refine the design, making it difficult to justify the costs within the project budget.

Time-Consuming Production Process: The time required to produce high-quality, detailed 3D prints can be significant, potentially delaying the technical design process, particularly if adjustments or redesigns are needed.

Challenges in Representing Complex Assemblies: 3D printing may struggle to accurately represent complex assemblies or systems with moving parts, leading to incomplete or oversimplified models that do not fully capture the intricacies of the technical design.

Difficulty in Iterating Design Changes: Iterating on technical designs using 3D printing can be cumbersome, as each change may require a new print, slowing down the ability to rapidly test and refine design solutions compared to digital modelling.

Integration with Digital Workflows: Integrating 3D-printed models with digital technical design workflows can be challenging, especially when ensuring that changes in the digital model are accurately reflected in the physical print, and vice versa.

Environmental Impact: The production of 3D-printed models can generate waste and consume resources, raising sustainability concerns, particularly if multiple prototypes are printed throughout the technical design phase.

Potential for Overreliance on Physical Models: Overreliance on 3D-printed models may reduce the use of advanced digital tools that offer more dynamic and detailed analysis capabilities, such as simulations or parametric design, which are critical in technical design.

Storage and Handling Issues: Physical 3D-printed models, especially large or complex ones, can be difficult to store and transport, which may create logistical challenges during the technical design phase, where frequent review and collaboration are required.

In the Technical Design phase, these barriers and challenges can limit the utility of 3D printing, necessitating careful consideration of its role and integration to ensure it complements rather than complicates the design process, while balancing cost, time, and precision requirements.

9.2.5.6 Stage 5: Manufacturing and Construction

Material Limitations: The range of materials that can be used in 3D printing is still limited, and many construction-grade materials, such as reinforced concrete or specific alloys, are not yet fully compatible with 3D printing technologies, limiting the application of printed components in actual construction.

Structural Integrity Concerns: 3D-printed components may not always match the strength and durability of traditionally manufactured parts, raising concerns about their structural integrity and long-term performance in construction projects.

High Initial Setup Costs: The investment required for industrial-scale 3D printers, along with the necessary materials and setup, can be substantial, making it difficult to justify within the tight budgets often associated with manufacturing and construction phases.

Production Speed: 3D printing, especially for large-scale components, can be time-consuming compared to traditional manufacturing methods, potentially slowing down the construction schedule and causing delays in project completion.

Challenges in Scaling Up: Scaling up 3D printing to meet the demands of large construction projects can be challenging, as the technology may struggle to produce large volumes of components quickly and consistently.

Integration with Traditional Construction Methods: Integrating 3D-printed components with traditionally manufactured or constructed elements can be complex, requiring precise coordination and potentially leading to compatibility issues on the construction site.

Quality Control Issues: Ensuring consistent quality in 3D-printed components can be challenging, particularly when producing complex geometries or when using less standardized materials, which can lead to variability and potential failures in the field.

Maintenance and Repair Difficulties: Repairing or modifying 3D-printed components can be more difficult than with traditionally manufactured parts, potentially leading to challenges in maintaining or retrofitting printed elements during or after construction.

Regulatory and Compliance Barriers: 3D printing in construction is still a relatively new technology, and there may be regulatory and compliance challenges, as building codes and standards may not fully address or recognize 3D-printed components, leading to delays or additional approvals.

Environmental Impact: The environmental impact of 3D printing, including energy consumption and material waste, can be significant, particularly if the process involves multiple iterations or if the printed components are not optimized for sustainability.

Training and Expertise Requirements: The use of 3D printing in construction requires specialized knowledge and skills, which may not be readily available on-site, necessitating additional training or the involvement of external experts, which can add to project costs and complexity.

In the Manufacturing and Construction phase, these barriers and challenges can limit the effectiveness and practicality of 3D printing, requiring careful planning, robust quality control measures, and strategic integration with traditional methods to ensure that the technology adds value without compromising project goals.

9.2.5.7 *Stage 6: Handover*

Difficulty in Updating As-Built Models: 3D-printed models may not easily reflect last-minute changes or adjustments made during construction, leading to potential discrepancies between the as-built 3D model and the actual completed building.

Limited Practical Use for Clients: Clients may find limited practical use for 3D-printed models post-handover, especially if they are not familiar with the technology or if the models do not add significant value beyond traditional documentation.

High Costs for Customization: Customizing 3D-printed models to accurately represent the final building can be expensive, potentially adding unnecessary costs during the handover phase without providing commensurate benefits to the client.

Storage and Preservation Issues: Physical 3D-printed models can be bulky and difficult to store, and they may degrade over time, leading to potential challenges in maintaining these models as part of the handover documentation.

Integration with Digital Documentation: Integrating 3D-printed models with digital handover documentation, such as BIM data, can be challenging, particularly if there are inconsistencies or if the model does not accurately capture all building details.

Potential for Overemphasis on Visuals: There is a risk that the focus on providing a tangible 3D model may detract from the completeness or quality of the digital documentation, which is often more useful for facility management and future maintenance.

Client Familiarity and Engagement: Clients may not be familiar with how to use or interpret 3D-printed models, leading to potential disengagement or confusion about the value these models add to the handover process.

Accuracy and Detail Limitations: The level of detail and accuracy that can be achieved with 3D printing may not be sufficient to capture all aspects of the completed project, leading to potential gaps in the handover deliverables.

Environmental Considerations: Producing physical 3D models as part of the handover can generate waste and consume resources, raising concerns about the environmental impact, especially if multiple iterations are needed.

Challenges in Demonstrating Functional Aspects: 3D-printed models are primarily visual and may not effectively demonstrate the functional aspects of the building, such as mechanical systems, energy performance, or maintenance requirements, which are critical for the client post-handover.

In the Handover phase, these barriers and challenges can limit the effectiveness and relevance of 3D-printed models, requiring careful consideration of their role in the overall documentation and how they can best complement digital tools and traditional methods to provide comprehensive and useful information to the client.

9.2.5.8 *Stage 7: Use*

Limited Practical Application Post-Construction: Once the building is in use, the practical applications of 3D printers diminish significantly, as most operational needs involve maintenance, repairs, and modifications that are better served by traditional methods.

High Costs for Maintenance and Updates: The ongoing cost of maintaining and operating 3D printers, including software updates, material supply, and potential repairs, may not be justified during the use phase, where the focus shifts to building operation and management.

Skill and Training Requirements: Facility management teams may lack the necessary expertise to operate 3D printers effectively, necessitating additional training or the hiring of specialized personnel, which can increase operational costs.

Integration Challenges with Facility Management: Integrating 3D-printed components or models into existing facility management systems can be complex, particularly if the models do not align with the digital data used for ongoing building management.

Limited Usefulness for Ongoing Maintenance: 3D printers may not be well-suited for producing replacement parts or components for building maintenance, especially if the required parts are large, complex, or made from materials that 3D printers cannot adequately replicate.

Environmental Impact of Continued Use: The ongoing use of 3D printers can have an environmental impact, particularly in terms of energy consumption and waste production, which may conflict with sustainability goals during the building's operational phase.

Potential for Obsolescence: As building technology evolves, 3D printers may quickly become outdated, requiring further investment in new equipment to remain functional and relevant during the use phase of the building.

Storage and Space Considerations: Maintaining 3D printers on-site requires dedicated space, which may be difficult to allocate in an occupied building where space is often at a premium.

Complexity in Producing Accurate As-Built Updates: If building modifications are needed, producing accurate as-built updates using 3D printing can be challenging, as the technology may not fully capture the nuances of changes made during the building's use.

Inconsistent Quality and Reliability: The quality and reliability of 3D-printed components can vary, especially if printers are not maintained properly or if there are inconsistencies in material supply, leading to potential issues in the ongoing use of printed parts.

In the Use phase, these barriers and challenges can limit the relevance and effectiveness of 3D printing, requiring careful evaluation of its role and potential benefits compared to traditional methods to ensure that it adds value without introducing unnecessary complexity or costs.

9.2.6 *Barriers to the Adoption of 3D Laser Scanners*

9.2.6.1 Stage 0: Strategic Definition

High Initial Costs: The acquisition and implementation of 3D laser scanning technology require significant upfront investment, which can be difficult to justify during the strategic definition phase when budgets are still being established.

Limited Perceived Value at Early Stages: The strategic definition phase is focused on high-level planning and goal-setting, where the detailed data provided by 3D laser scanners may not be immediately necessary, leading to questions about the cost-benefit ratio at this stage.

Technical Expertise Requirements: Effective use of 3D laser scanners requires specialized technical skills and knowledge, which may not be available within the project team during the strategic phase, potentially necessitating external consultants or additional training.

Integration with Strategic Planning Tools: Integrating 3D laser scanning data with traditional strategic planning tools and processes can be challenging, leading to potential inefficiencies or delays if the technology is not seamlessly aligned with existing workflows.

Complexity in Data Interpretation: The data generated by 3D laser scanners can be highly detailed and complex, which may overwhelm stakeholders or decision-makers who are not familiar with interpreting such data during the strategic definition phase.

Overreliance on Technology: There is a risk of overemphasis on the capabilities of 3D laser scanners, potentially leading to the neglect of other important strategic considerations, such as stakeholder engagement or broader project objectives.

Regulatory and Compliance Concerns: The use of 3D laser scanning technology may raise regulatory or compliance issues, particularly if the scanning involves sensitive sites or areas with specific data protection requirements, adding complexity to the strategic planning process.

Time-Consuming Data Processing: Processing and analysing the large volumes of data generated by 3D laser scanners can be time-consuming, potentially delaying strategic decision-making and hindering the overall planning timeline.

Dependence on High-Quality Input Data: The effectiveness of 3D laser scanning is dependent on the quality of the input data, and any inaccuracies or limitations in the initial scanning process can lead to flawed strategic decisions based on incomplete or incorrect information.

Environmental and Site Accessibility Limitations: Environmental factors or site accessibility issues may limit the effectiveness of 3D laser scanning, particularly in challenging terrains or areas with limited access, potentially reducing the technology's value during the strategic phase.

In the Strategic Definition phase, these barriers and challenges can limit the perceived and practical value of 3D laser scanning, requiring careful consideration of its application, cost implications, and alignment with broader strategic goals to ensure it contributes positively to the project's early planning stages.

9.2.6.2 Stage 1: Preparation and Brief

High Initial Costs: The purchase and setup of 3D laser scanning equipment are costly, which can be difficult to justify during the preparation and brief phase when budgets are still being finalized and resources are allocated.

Technical Expertise Requirements: Utilizing 3D laser scanners effectively requires specialized knowledge and skills, which may not be readily available within the project team at this early stage, necessitating additional training or hiring of experts.

Complexity in Data Interpretation: The data generated by 3D laser scanners are often complex and detailed, requiring significant effort to interpret and integrate into the project brief, potentially leading to delays or misunderstandings if not managed properly.

Data Overload: The vast amount of data captured by 3D laser scanners can be overwhelming, making it challenging for project teams to focus on the most relevant information needed for the preparation and brief phase.

Integration with Existing Processes: Incorporating 3D laser scanning data into traditional preparation and brief workflows can be challenging, particularly if existing processes are not designed to handle such detailed digital data.

Time-Consuming Data Processing: Processing the high-resolution data from 3D laser scanners can be time-consuming, potentially delaying the completion of the project brief and the overall project timeline.

Limited Applicability for Conceptual Work: During the preparation and brief phase, the focus is often on high-level concepts and objectives, where the detailed, point-by-point accuracy of 3D laser scanning may be less relevant, leading to underutilization of the technology.

Potential for Overreliance on Technology: Relying too heavily on 3D laser scanning data can overshadow other important aspects of the project brief, such as stakeholder input, client requirements, and broader strategic goals.

Environmental and Site Limitations: The effectiveness of 3D laser scanners can be limited by environmental conditions, such as lighting, weather, or obstructions on the site, which may reduce the accuracy or completeness of the data collected.

Regulatory and Privacy Concerns: Using 3D laser scanners, especially in populated or sensitive areas, may raise regulatory and privacy issues, complicating the preparation and brief process if additional permissions or protections are required.

In the Preparation and Brief phase, these barriers and challenges can hinder the effective use of 3D laser scanners, requiring careful consideration of their role, integration with existing processes, and the management of data to ensure they contribute effectively to the project's early development.

9.2.6.3 Stage 2: Concept Design

High Costs for Early-Stage Use: The expense of acquiring and operating 3D laser scanners may be difficult to justify during the concept design phase, where budgets are still being allocated, and the need for detailed data may not yet be critical.

Overly Detailed Data for Conceptual Work: 3D laser scanners provide highly detailed and precise data, which may be excessive for the conceptual nature of this phase, potentially leading to unnecessary complexity in the design process.

Technical Expertise Requirements: Effective use of 3D laser scanners requires specialized technical skills, which may not be readily available within the design team during the concept phase, leading to potential delays or the need for external support.

Integration Challenges with Creative Design Tools: Integrating 3D laser scanning data with creative design tools can be challenging, particularly if the design team relies on software that is not fully compatible with the detailed data produced by laser scanners.

Time-Consuming Data Processing: The processing and interpretation of 3D laser scanning data can be time-consuming, potentially slowing down the iterative design process that is essential in the concept design phase.

Risk of Design Fixation: The precise nature of 3D laser scanning data might lead designers to fixate on existing site conditions, potentially limiting creativity and the exploration of innovative design solutions.

Environmental and Site Accessibility Issues: The effectiveness of 3D laser scanners can be limited by environmental factors such as poor lighting or restricted site access, which may result in incomplete or inaccurate data that hampers the concept design process.

Potential Overreliance on Technology: There is a risk that the design team may over-rely on 3D laser scanning data, neglecting other important aspects of concept design, such as stakeholder input, client aspirations, and broader project goals.

Challenges in Visualizing Abstract Concepts: 3D laser scanners are excellent for capturing existing conditions, but they may not be as effective in helping designers visualize abstract or conceptual ideas, which are critical during the early stages of design.

Data Management and Storage: The large volumes of data generated by 3D laser scanners require robust management and storage solutions, which can be burdensome and may divert attention from more creative aspects of the concept design phase.

In the Concept Design phase, these barriers and challenges can limit the practical benefits of 3D laser scanners, requiring careful consideration of their role and integration to ensure they enhance rather than hinder the creative and exploratory nature of this stage.

9.2.6.4 *Stage 3: Spatial Coordination*

High Initial Costs: The expense of purchasing and operating 3D laser scanners can be significant, potentially making it challenging to justify their use solely for spatial coordination, especially in projects with tight budgets.

Complexity of Data Integration: Integrating the detailed data from 3D laser scanners into BIM or other spatial coordination tools can be complex, requiring significant time and technical expertise to ensure compatibility and accuracy.

Data Overload: The high level of detail provided by 3D laser scanners can lead to data overload, making it difficult for the project team to focus on the most relevant information needed for effective spatial coordination.

Time-Consuming Data Processing: Processing the extensive data collected by 3D laser scanners can be time-consuming, potentially delaying the spatial coordination process and impacting the overall project timeline.

Technical Expertise Requirements: Using 3D laser scanners effectively requires specialized skills in data interpretation and software integration, which may not be readily available within the project team, necessitating additional training or external support.

Potential for Inaccuracies in Complex Environments: In environments with complex geometries or obstructed views, 3D laser scanners may struggle to capture accurate data, leading to potential inaccuracies in the spatial coordination model.

Environmental and Site Limitations: Factors such as lighting conditions, reflective surfaces, or physical obstructions can affect the accuracy and completeness of the data collected by 3D laser scanners, which can complicate spatial coordination efforts.

Difficulty in Managing Large Data Sets: The large data sets generated by 3D laser scanners require robust data management solutions, which can be resource-intensive and may distract from other critical aspects of spatial coordination.

Challenges in Iterative Coordination: Iterative spatial coordination processes may require multiple rounds of data capture and analysis, which can be cumbersome and time-consuming when using 3D laser scanners, particularly if frequent updates are needed.

Overreliance on Technology: There is a risk that project teams may become overly reliant on 3D laser scanning data, potentially overlooking the importance of manual checks, on-site inspections, and other traditional spatial coordination methods.

In the Spatial Coordination phase, these barriers and challenges can limit the effectiveness of 3D laser scanners, requiring careful planning, skilled operation, and strategic integration to ensure that the technology enhances coordination without introducing unnecessary complexity or delays.

9.2.6.5 *Stage 4: Technical Design*

High Implementation Costs: The cost of acquiring, operating, and maintaining 3D laser scanners can be substantial, making it difficult to justify their use during the technical design phase, especially in projects with constrained budgets.

Complex Data Integration: Integrating the highly detailed data from 3D laser scanners into technical design software can be challenging, particularly if the software does not seamlessly support the data formats or if the design team lacks experience in managing such data.

Time-Consuming Data Processing: The processing and refinement of the extensive data captured by 3D laser scanners can be time-consuming, potentially slowing down the technical design process and affecting project timelines.

Overwhelming Level of Detail: The high level of detail provided by 3D laser scanners may be excessive for certain aspects of technical design, leading to potential distractions or inefficiencies as the design team sifts through large volumes of data.

Technical Expertise Requirements: Effective use of 3D laser scanning data in technical design requires specialized knowledge and skills, which may not be readily available within the project team, necessitating additional training or reliance on external experts.

Potential for Data Inaccuracies: In environments with complex geometries, reflective surfaces, or obstructions, 3D laser scanners may produce inaccurate or incomplete data, which can compromise the precision required in technical design.

Challenges in Capturing Dynamic Elements: 3D laser scanners are excellent for capturing static objects but may struggle with dynamic elements or systems, such as moving parts or elements that change over time, which can limit their usefulness in technical design.

Dependence on Optimal Conditions: The accuracy and reliability of 3D laser scanning data can be affected by environmental conditions, such as lighting, weather, or physical obstructions, potentially leading to inaccuracies in the technical design.

Data Management and Storage Issues: The large data sets generated by 3D laser scanners require robust data management and storage solutions, which can add to the complexity and cost of the technical design phase.

Risk of Overreliance on Technology: There is a risk that design teams may become overly reliant on 3D laser scanning data, potentially neglecting traditional design methods or critical manual checks that are essential for ensuring the accuracy and feasibility of the design.

In the Technical Design phase, these barriers and challenges can limit the effectiveness and practicality of 3D laser scanners, requiring careful consideration of their role, the integration of complementary technologies, and the development of strategies to manage the data and costs associated with their use.

9.2.6.6 Stage 5: Manufacturing and Construction

High Equipment and Operation Costs: The purchase, operation, and maintenance of 3D laser scanners are expensive, making it challenging to justify their use in the manufacturing and construction phases, especially in budget-conscious projects.

Technical Expertise Requirements: Operating 3D laser scanners and interpreting the data they produce requires specialized skills, which may not be readily available on construction sites, necessitating additional training or the hiring of experts.

Complex Integration with Construction Workflows: Integrating 3D laser scanning data into existing construction workflows can be difficult, particularly if the data formats are incompatible with other construction software or tools used on-site.

Time-Consuming Data Processing: The detailed data captured by 3D laser scanners often requires significant processing time, which can delay construction activities, especially if rapid decision-making or adjustments are needed on-site.

Environmental and Site Limitations: The accuracy of 3D laser scanning can be affected by environmental factors such as lighting, weather conditions, and physical obstructions, leading to potential inaccuracies that can complicate the construction process.

Challenges in Capturing Moving Elements: 3D laser scanners are well-suited for capturing static objects but may struggle to accurately capture or monitor dynamic elements, such as moving machinery or ongoing construction activities, limiting their utility in active construction environments.

Data Overload: The vast amount of data generated by 3D laser scanners can overwhelm construction teams, making it difficult to focus on the most critical information needed for decision-making and quality control.

Potential for Data Inaccuracies: In complex or crowded construction sites, 3D laser scanners may produce inaccurate or incomplete data, which can lead to errors in the construction process if relied upon too heavily without verification.

Logistical Challenges on Construction Sites: Deploying 3D laser scanners on busy construction sites can be logistically challenging, as they require stable platforms and unobstructed views, which may not always be feasible in dynamic construction environments.

Dependence on Digital Infrastructure: Effective use of 3D laser scanners requires reliable digital infrastructure, including high-speed data transfer and storage solutions, which may not be fully available on all construction sites, especially in remote or less developed areas.

In the Manufacturing and Construction phase, these barriers and challenges can limit the practical benefits of 3D laser scanners, necessitating careful planning, skilled operation, and strategic integration with other construction tools and methods to ensure they enhance rather than hinder the construction process.

9.2.6.7 Stage 6: Handover

High Costs with Limited Return: The costs associated with using 3D laser scanners during the handover phase, including equipment, operation, and data processing, may be high relative to the perceived benefits, especially when traditional methods could suffice.

Data Overload for Clients: The detailed data generated by 3D laser scanners can be overwhelming for clients, particularly if they are not familiar with interpreting such data, leading to challenges in understanding and utilizing the **information** effectively.

Integration with Facility Management Systems: Integrating the 3D laser scanning data into existing facility management systems can be complex, particularly if the data formats are not fully compatible or if the client's systems are not equipped to handle such detailed information.

Technical Expertise Requirements: The handover of 3D laser scanning data may require the client's team to have specialized technical knowledge to interpret and manage the data effectively, which may necessitate **additional** training or support.

Challenges in Capturing Final As-Built Conditions: If construction activities continue until the final stages, capturing the exact as-built conditions with 3D laser scanners can be challenging, leading to potential discrepancies between the scanned data and the actual final state of the building.

Potential for Overreliance on Technology: There is a risk that project teams may over-rely on 3D laser scanning data during handover, potentially overlooking the importance of more traditional documentation methods that clients may find easier to use and understand.

Data Accuracy and Completeness Issues: Any inaccuracies or gaps in the 3D laser scanning data collected during the final stages of construction can result in incomplete or incorrect as-built records, which may cause issues in future building maintenance or modifications.

Storage and Management of Large Data Sets: The large data sets generated by 3D laser scanners require significant storage and management solutions, which can be burdensome for both the project team and the client during the handover process.

Potential for Limited Client Engagement: Clients who are not technologically inclined may struggle to engage with or fully appreciate the value of 3D laser scanning data, leading to a disconnect between the project team's efforts and the client's ability to utilize the information.

Environmental and Site Constraints: If the building is in use during the handover phase, environmental factors or site accessibility issues may limit the effectiveness of 3D laser scanning, leading to potential gaps in the final documentation.

In the Handover phase, these barriers and challenges can limit the effectiveness and practicality of using 3D laser scanners, requiring careful planning, clear communication with the client, and strategic integration to ensure the technology adds value to the final project deliverables.

9.2.6.8 Stage 7: Use

High Maintenance and Operational Costs: The ongoing costs associated with maintaining and operating 3D laser scanners can be significant, making it challenging to justify their continued use during the building's operational phase.

Limited Practical Applications Post-Construction: Once the building is in use, the need for 3D laser scanning may diminish, as the technology is more suited to capturing as-built conditions rather than ongoing operational tasks, which may limit its practical applications.

Complex Data Management: Managing and storing the vast amounts of data generated by 3D laser scanners can be cumbersome, requiring robust digital infrastructure and data management practices, which may be difficult to maintain over the building's lifecycle.

Specialized Skill Requirements: Effective use of 3D laser scanners in the operational phase requires specialized technical skills, which may not be readily available within the facility management team, necessitating additional training or external support.

Integration with Facility Management Systems: Integrating 3D laser scanning data with existing facility management systems can be complex, particularly if those systems are not designed to handle such detailed and large data sets, potentially leading to inefficiencies.

Potential for Data Inaccuracy Over Time: As the building is used and undergoes modifications, the original 3D laser scanning data may become outdated, leading to inaccuracies if relied upon too heavily for maintenance or future renovations.

Environmental and Operational Constraints: The effectiveness of 3D laser scanning during the use phase can be limited by environmental conditions or the presence of occupants, which may restrict access or impact the accuracy of the scans.

Security and Privacy Concerns: Using 3D laser scanners in occupied buildings may raise security and privacy concerns, particularly in sensitive areas where detailed scans could capture confidential or proprietary information.

Limited Usefulness for Routine Maintenance: For day-to-day maintenance tasks, 3D laser scanners may offer limited benefits compared to more traditional methods, as the detailed scans may not be necessary for routine inspections or repairs.

Risk of Overreliance on Technology: There is a risk that facility managers may become overly reliant on 3D laser scanning data, potentially neglecting other important aspects of building maintenance that require manual inspection and hands-on management.

In the Use phase, these barriers and challenges can limit the practicality and effectiveness of 3D laser scanners, requiring careful consideration of their role in ongoing building operations and ensuring that they are integrated in a way that complements, rather than complicates, facility management practices

9.3 Conclusions

The construction industry stands at a pivotal crossroads, where the adoption of digital technologies is no longer a luxury but a necessity for sustainability and competitiveness. The barriers to this adoption – cultural resistance, technical challenges, financial constraints, lack of expertise, and regulatory hurdles – while significant, are not insurmountable. This conclusion synthesizes the insights from the chapter, highlighting the importance of overcoming these barriers and proposing actionable strategies for industry-wide digital transformation.

Cultural Resistance and Organizational Change: One of the most formidable barriers to digital adoption in construction is the industry's deeply ingrained traditionalism. Overcoming this requires a paradigm shift in organizational culture, where leadership plays a critical role in fostering openness to innovation. By championing digital technologies and demonstrating their value, leaders can cultivate a culture that embraces change rather than fearing it.

Technical Challenges and Interoperability: The fragmentation of digital solutions and the lack of interoperability pose significant technical challenges. Addressing these requires a concerted effort towards standardization and the development of integrated platforms that facilitate seamless data exchange. Collaboration among technology providers, alongside investment in R&D, can pave the way for more cohesive and user-friendly digital ecosystems.

Financial Constraints and Investment: The high cost of digital technologies and the uncertainty regarding their ROI deter investment. To mitigate these financial constraints, the industry needs innovative financing models and incentives that lower the entry barrier for digital adoption. Public–private partnerships and government subsidies can play a pivotal role in easing the financial burden, making digital technologies more accessible to all industry players.

Skills Gap and Education: The digital transformation of construction demands a workforce skilled in new technologies. Bridging the skills gap requires comprehensive education and training programmes focused on digital literacy and technical competencies. Universities, vocational schools, and industry associations must collaborate to update curricula and offer continuous learning opportunities that align with the evolving technological landscape.

Regulatory Support and Policy Frameworks: The lack of supportive regulatory frameworks hinders digital adoption. Governments and regulatory bodies must develop policies that encourage the use of digital technologies, including standards for data security, privacy, and interoperability. Clear guidelines and standards can alleviate concerns and foster an environment conducive to digital innovation.

Collaboration and Community Engagement: The fragmented nature of the construction industry often leads to siloed efforts in digital adoption. Building a collaborative community that shares knowledge, resources, and best practices is essential. Industry-wide forums, workshops, and consortia can facilitate the exchange of ideas and foster partnerships that accelerate digital transformation.

Innovation and Future Technologies: Looking ahead, the continuous evolution of digital technologies promises to further revolutionize the construction industry. Emerging technologies such as artificial intelligence, AR, and blockchain offer new opportunities for efficiency, transparency, and sustainability. Embracing these innovations requires an ecosystem that supports experimentation and risk-taking.

In conclusion, overcoming the barriers to digital adoption in construction is a multifaceted challenge that demands a collaborative, strategic approach. By addressing cultural, technical, financial, and regulatory challenges, and by investing in education and innovation, the construction industry can unlock the full potential of digital technologies. This transformation will not only improve project outcomes and operational efficiency but also contribute to the industry's long-term resilience and growth. As we move forward, it is imperative for all stakeholders – industry leaders, policymakers, educators, and technology providers – to unite in their efforts to drive digital adoption, shaping a future where digital construction is not just a possibility but a reality.

Chapter 10

Future Directions

10.1 Introduction

This chapter aims to explore the future directions of digital technologies in the engineering and construction sectors, building upon the comprehensive foundation laid out in the preceding chapters of 'Digital Technologies for Engineering Construction Information Productivity.' As we navigate through the transformative landscapes of construction and engineering, this concluding chapter not only synthesizes the insights and advancements discussed but also projects the trajectory of digital technologies. It serves as a signpost, guiding readers through potential innovations, challenges, and opportunities that lie ahead in integrating digital technologies within the construction industry.

10.2 A Perspective

In the initial chapters, the book lays a robust foundation by introducing the critical global challenges and opportunities within the construction industry, emphasizing the significance of zero-energy and carbon buildings alongside strategies for affordable housing. This sets the stage for understanding the necessity of integrating digital technologies in addressing these challenges. It then transitions into a deep dive into the realm of digital construction technologies, elaborating on the transformative potential of robots, D-laser scanners, 3D printers, blockchains, and drones. This exploration is not just theoretical; it is rich with practical implications, showcasing how these technologies revolutionize construction processes, enhance precision, and improve efficiency.

10.2.1 Building Information Modelling (BIM)

BIM receives a dedicated focus, unravelling its complexities and demonstrating its indispensable role in the digital transformation of construction. Through a detailed examination of BIM software systems, the narrative clarifies how BIM serves as a pivotal tool for decision-making and efficiency throughout a building's lifecycle. The discourse on BIM extends into discussions on common concepts, system integration, and interoperability, highlighting the essential elements for seamless operation and collaboration within the construction industry. These sections emphasize the critical nature of data exchange and the integration of various digital technologies to streamline construction projects.

Practical applications of digital technologies in construction are extensively covered, revealing how BIM and other digital tools can be applied across various stages of construction

DOI: 10.1201/9781003476740-10

projects. From design and scheduling to cost estimation and clash detection, the book illustrates the comprehensive capabilities of BIM, showcasing its impact on sustainability, retrofitting, and renovation projects. The narrative then shifts towards the innovative realm of cloud-based BIM, discussing the advent of mobile BIM and the importance of common data environments for effective information management. This section underscores the transformational impact of cloud platforms in facilitating collaboration and accessibility of BIM data across project team.

10.2.2 Concept of Information Exchange

The 'Common Concepts, System Integration & Interoperability' section of the book delves into the intricate dynamics of integrating various digital technologies within the construction industry. It begins by elucidating common BIM concepts, such as the BIM maturity models, which describe the levels of BIM adoption within organizations, ranging from basic 2D CAD drawings to fully integrated and collaborative processes. The BIM level of development (LOD) framework is also introduced, providing a detailed system to specify the precision and content of BIM elements at different stages of the project lifecycle.

Further exploration into nD modelling reveals how BIM extends beyond mere visual representation, incorporating time (4D) for scheduling, cost (5D), environmental considerations (6D), and lifecycle management (7D). This multidimensional approach allows for a holistic view of projects, facilitating better decision-making and efficiency.

The discussion on the BIM team and the BIM periodic table highlights the importance of collaboration and the diverse skill sets required to leverage BIM technology fully. The BIM team concept underscores the collaborative framework necessary for successful BIM implementation, involving architects, engineers, contractors, and other stakeholders. The BIM periodic table, meanwhile, serves as a metaphorical representation of the essential elements and standards required for effective BIM practices, emphasizing the need for a comprehensive understanding of these components to achieve interoperability and seamless integration.

System integration is presented as a critical challenge and opportunity within the construction industry. It involves the technical and organizational process of linking different computing systems and software applications physically or functionally, to act as a coordinated whole. The book stresses the significance of system integration in ensuring that various digital tools and platforms used in construction projects can communicate and work together efficiently, thereby reducing data silos and enhancing project outcomes.

Interoperability, closely related to system integration, is discussed in terms of its importance in allowing different information technology systems and software applications to communicate, exchange data, and use the information that has been exchanged. The text highlights that interoperability is crucial for the effective use of BIM and other digital construction technologies across different phases of the construction process. It facilitates the seamless sharing of data and information, ensuring that all stakeholders can access and utilize the data they need in a compatible format, thereby enhancing collaboration and efficiency.

The chapter also addresses the challenges to achieving system integration and interoperability, such as the lack of standardized data formats and protocols, the proprietary nature of many software solutions, and the cultural and organizational barriers within the construction industry. It advocates for the adoption of open standards and collaborative approaches to overcome these challenges.

In conclusion, the 'Common Concepts, System Integration & Interoperability' section articulates a vision where fully integrated and interoperable digital construction technologies become a cornerstone of the construction industry. This integration promises to streamline project workflows, enhance communication among stakeholders, and lead to more sustainable, efficient, and cost-effective construction projects. The text calls for ongoing collaboration, standardization efforts, and a willingness to embrace new technologies and processes as essential steps towards realizing this vision.

10.2.3 Digital Construction Applications

The chapter on 'Digital Construction Applications' delves into the multifaceted ways in which digital technologies, particularly BIM, are applied within the construction industry to revolutionize traditional processes and enhance project outcomes. It begins by outlining the significance of BIM in the design phase, where it serves as a collaborative tool that enables architects, engineers, and construction professionals to create and manipulate digital representations of physical and functional characteristics of places. This collaborative environment fosters improved decision-making, enhanced accuracy, and a reduction in design errors, leading to a more efficient planning stage.

The application of BIM extends beyond design into the generation of detailed project schedules and timelines. By integrating BIM with project management software, stakeholders can visualize the construction process in a simulated environment, allowing for better planning and allocation of resources. This integration significantly impacts the project's efficiency by enabling precise scheduling, reducing the likelihood of conflicts, and facilitating smoother project execution.

Another critical application discussed is the use of BIM for quantity take-offs and cost estimation. Traditional methods of estimating costs and quantities are often time-consuming and prone to errors. BIM automates these processes, providing accurate and real-time data that can drastically reduce the time required for quantity take-offs and improve the reliability of cost estimates. This accuracy is crucial for budgeting and financial planning, ensuring projects stay within budget and financial resources are allocated efficiently.

BIM's role in clash detection is highlighted as a game-changer in preventing construction errors before they occur. By identifying and resolving conflicts between different design elements in the virtual model, BIM prevents costly and time-consuming rework. This proactive approach to identifying and solving problems can lead to significant savings in time and money, enhancing overall project efficiency.

Furthermore, the chapter explores BIM's contributions to project scheduling and programme management. By allowing for the creation of more detailed and accurate schedules, BIM facilitates a more streamlined approach to managing construction projects. This capability is particularly beneficial in large-scale projects with complex timelines, where coordination between different teams and tasks is critical.

Sustainability is another area where BIM's applications are thoroughly examined. Through the integration of sustainable design principles and the analysis of energy performance, BIM enables the construction of buildings that are more environmentally friendly and energy-efficient. This application of BIM is crucial for meeting global sustainability goals and reducing the construction industry's environmental impact.

The chapter also addresses BIM's role in retrofitting, renovation, and refurbishment projects, where its ability to model existing conditions and visualize proposed changes offers significant advantages. This capability is particularly valuable in projects involving historical preservation, where accuracy and sensitivity to existing structures are paramount.

In the realm of structural engineering, BIM's applications are shown to extend to the detailed analysis and design of building structures, enabling more efficient material use and innovative architectural forms. Additionally, the chapter underscores the importance of collaboration facilitated by BIM, enhancing communication and coordination among all project stakeholders, which is fundamental to the successful delivery of construction projects.

Lastly, the integration of BIM with other software tools like Revit, Power BI, and various Autodesk products is discussed, showcasing how these integrations can optimize productivity and project outcomes. These applications illustrate the dynamic and evolving nature of digital construction technologies, highlighting the vast potential of BIM to transform every aspect of the construction process, from initial design to final execution.

10.2.4 Cloud BIM

The chapter on 'cloud-based BIM' delves into the transformative shift towards leveraging cloud computing within the realm of BIM, marking a significant evolution in how construction projects are managed, collaborated on, and executed. This paradigm shift to cloud-based BIM platforms facilitates unprecedented levels of accessibility, efficiency, and collaboration, breaking down the traditional barriers of time and location that have historically constrained project stakeholders.

Cloud-based BIM represents a nexus of digital collaboration, where project data and BIM models are stored in a centralized cloud repository. This enables project teams, regardless of their geographical locations, to access, update, and manage project information in real-time. Such a model promotes a more integrated project delivery approach, enhancing communication among architects, engineers, contractors, and clients. It ensures that all stakeholders are working from the latest information, thereby reducing errors, avoiding rework, and improving overall project outcomes.

Moreover, cloud-based BIM extends beyond mere data storage to include a suite of tools and applications designed to support various aspects of construction management, including document management, field management, design collaboration, model coordination, project management, and asset management. These tools are tailored to streamline the workflows, facilitate seamless data exchange, and enhance decision-making processes across the project lifecycle.

One of the key features of cloud-based BIM is its support for mobile BIM applications. This mobility allows stakeholders to access BIM models and project data on-the-go, using smartphones or tablets. This capability is crucial for on-site workers and project managers, enabling them to review designs, perform inspections, and update project statuses directly from the construction site, thereby enhancing productivity and decision-making speed.

Cloud-based BIM also emphasizes the importance of CDEs for information management. A CDE provides a single source of truth for all project information, ensuring that data are consistently managed and maintained throughout the project lifecycle. This approach minimizes data silos, ensures consistency in information, and supports regulatory compliance by maintaining a comprehensive audit trail of decisions and changes.

The discussion on cloud-based BIM wouldn't be complete without mentioning BIM 360, Autodesk's suite of tools designed to support cloud-based BIM workflows. BIM 360 encompasses a range of functionalities, from document and field management to design collaboration and model coordination, all integrated within a cloud platform. It exemplifies how cloud-based BIM can enhance project management, improve collaboration, and accelerate project delivery.

In conclusion, cloud-based BIM represents a significant leap forward in the construction industry's ongoing digital transformation. It offers a powerful platform for enhancing collaboration, improving efficiency, and driving better project outcomes. As the industry continues to embrace digital technologies, cloud-based BIM stands out as a critical tool in the arsenal, enabling construction professionals to meet the challenges of modern construction projects more effectively.

10.2.5 Information Management

The chapter on 'Information Management' within the context of digital construction technologies delves into the intricacies of managing vast amounts of data generated and utilized throughout the lifecycle of construction projects. This chapter highlights the critical importance of efficient information management systems and strategies in optimizing the productivity and effectiveness of construction projects. It explores the fundamental processes, technologies, and frameworks that enable construction professionals to capture, store, manage, and disseminate project information in a timely and accurate manner.

In the realm of construction, information management encompasses a broad spectrum of data types, including design documents, contracts, schedules, cost estimates, and project communications. The chapter emphasizes the necessity of establishing a CDE as a centralized repository for project information, facilitating access and collaboration among all stakeholders. The CDE serves as the backbone of information management, ensuring that all project participants have access to the latest, most accurate project data, thus minimizing errors and inconsistencies.

The chapter further explores the role of BIM in enhancing information management practices. BIM, beyond being a design tool, is portrayed as a powerful platform for information management, integrating various data types into a cohesive, interactive model that evolves throughout the project lifecycle. BIM's capability to link graphical data with non-graphical data allows for a more comprehensive understanding of the project, supporting decision-making processes, and improving communication among stakeholders.

Moreover, the chapter addresses the challenges of information management in construction, such as data interoperability, the need for standardized data formats, and the importance of cybersecurity measures to protect sensitive project information. It discusses the adoption of international standards, such as ISO 19650, which provides guidelines for managing information over the whole life cycle of a built asset using BIM.

The future of information management in construction is also a key focus, with the chapter predicting the increased use of artificial intelligence (AI) and machine learning algorithms to analyse project data, predict outcomes, and automate routine tasks. The integration of Internet of Things (IoT) devices and sensors on construction sites is anticipated to provide real-time data feeds, further enriching project information management systems and enabling more dynamic and responsive project management practices.

In conclusion, the chapter on 'Information Management' underscores the transformative potential of advanced information management strategies and technologies in the construction industry. By adopting effective information management practices, construction projects can achieve higher levels of efficiency, accuracy, and collaboration, paving the way for more successful project outcomes and the realization of innovative construction methodologies.

10.2.6 *Benefits of DCT*

The benefits of digital construction technologies encompass a wide array of improvements in the efficiency, sustainability, and quality of construction projects. These technologies, including BIM, 3D printing, drones, and augmented reality (AR)/virtual reality (VR), offer transformative potential that significantly impacts project delivery, cost, and time efficiency.

BIM stands at the forefront, serving as a pivotal tool that enhances collaboration among stakeholders by providing a digital representation of a building's physical and functional characteristics. BIM facilitates better decision-making throughout a building's life cycle, from the earliest concepts to demolition, by allowing project teams to visualize projects in a simulated environment. This visualization aids in identifying potential design, construction, or operational issues before they occur, thereby reducing errors, saving time, and minimizing waste.

3D printing in construction introduces a revolutionary approach to building fabrication. It enables the rapid prototyping and production of complex components, often at a lower cost and with less material waste than traditional construction methods. This technology not only accelerates the construction process but also opens up new possibilities for innovative design and construction techniques, making it possible to create structures that were once considered impractical or too expensive to build.

The use of drones has also become increasingly prevalent in the construction industry. Drones provide a unique vantage point, allowing for aerial surveys, progress tracking, and inspection of hard-to-reach areas without the need for costly and time-consuming manual labour. This capability enhances site safety by reducing the need for workers to perform dangerous tasks in elevated or otherwise hazardous conditions.

AR and VR technologies offer immersive experiences that can significantly enhance planning, design, and training processes. AR overlays digital information onto the real world, aiding in site planning and the visualization of future projects on existing landscapes. VR, on the other hand, creates a completely immersive simulation environment, ideal for stakeholder engagement, design verification, and worker training without the physical constraints of the real world.

Moreover, digital construction technologies contribute to sustainability in the construction industry. By optimizing resource use, reducing waste, and improving energy efficiency, these technologies support the creation of greener buildings and infrastructure. For instance, BIM's ability to simulate energy performance helps in designing buildings that consume less energy, while 3D printing minimizes material waste by using exactly the amount of material required for each component.

The adoption of digital construction technologies also brings about significant improvements in project management and communication. Cloud-based platforms and collaborative tools enable real-time sharing of project information, making it easier for teams to stay informed and make timely decisions. This improved communication reduces the likelihood of misunderstandings and delays, further enhancing project efficiency.

In conclusion, the benefits of digital construction technologies are manifold, touching upon virtually every aspect of the construction process. From the initial design phases to the final construction and beyond, these technologies offer a pathway to more efficient, sustainable, and innovative construction practices. As the industry continues to evolve, the adoption and integration of these technologies will play a critical role in shaping the future of construction.

10.2.7 Barriers of DCT

The adoption of digital construction technologies, while promising significant improvements in efficiency, sustainability, and overall project delivery, is not without its challenges. The barriers to the widespread integration of these technologies into the construction industry are multi-faceted, stemming from a variety of sources ranging from cultural resistance to technical and financial constraints.

One of the primary hurdles is the resistance to change. The construction sector, known for its traditional practices, often exhibits a hesitancy to adopt new technologies. This resistance is not merely a matter of preference but is deeply rooted in the industry's culture, where established methods are trusted over novel, unproven solutions. The reluctance is further compounded by a lack of understanding and awareness of the benefits these digital tools can bring, making the case for change harder to articulate to stakeholders accustomed to conventional practices.

Financial constraints also play a critical role in hampering the adoption of digital technologies. The initial investment required for implementing systems like BIM, 3D printing, or drone technology can be substantial. Small- to medium-sized enterprises (SMEs), which make up a significant portion of the industry, may find these costs prohibitive, especially without clear, immediate returns on investment. This financial barrier is exacerbated by the current economic models in the construction sector, which do not always facilitate or reward innovation and investment in technology.

Technical challenges also present significant barriers. The lack of interoperability between different digital construction tools and platforms can hinder their effective use. Many construction projects involve multiple stakeholders using different systems and software, leading to issues in data compatibility and exchange. This lack of standardization can result in inefficiencies and errors, negating some of the benefits digital technologies aim to provide.

The skills gap within the industry further complicates the adoption of digital technologies. There is a pressing need for training and education to equip the current and future workforce with the necessary skills to effectively use these tools. However, the pace of technological advancement often outstrips the development of training programmes, leaving a gap between the capabilities of the workforce and the potential offered by digital construction technologies.

Finally, regulatory and legal issues can act as barriers to the adoption of digital technologies. The construction industry is heavily regulated, and new technologies often outpace the development of relevant standards and regulations. This can create uncertainty regarding compliance, liability, and insurance, discouraging companies from investing in new digital tools.

Overcoming these barriers requires a concerted effort from all stakeholders in the construction industry. This includes increasing awareness and education about the benefits of digital technologies, developing financial models that support innovation, standardizing data and interoperability protocols, investing in workforce training, and updating regulations to accommodate new technologies. Only through addressing these challenges can the construction industry fully leverage the potential of digital construction technologies to transform its practices and outputs.

10.3 Limitations

The exploration within 'Digital Technologies for Engineering Construction Information Productivity' offers a comprehensive look at the intersection of construction and cutting-edge digital tools. However, the limitations inherent to this exploration stem from several key areas:

Technological Evolution: The rapid pace of technological innovation means that some of the digital tools and methodologies discussed may soon be superseded by more advanced solutions. The book captures a moment in time, and the continuous evolution of technology could lead to new tools that are not covered within its pages.

Geographical Variability: The book's discussions may not fully encapsulate the geographical differences in technology adoption and regulatory environments. Construction practices and the integration of digital technologies vary significantly across regions due to differences in economic conditions, regulatory frameworks, and cultural attitudes towards innovation.

Human and Organizational Factors: While focusing on the technological aspects of digital construction, the book may not delve deeply into the human and organizational factors that play a critical role in the adoption and effective use of these technologies. Issues such as workforce training, change management, and the digital divide among firms can significantly impact the success of technology implementation in construction projects.

Sustainability and Ethical Considerations: As the construction industry moves towards greater sustainability, the environmental and ethical implications of digital technologies require more attention. The book could further explore how digital tools can contribute to sustainable construction practices and the ethical considerations in automating construction processes, impacting labour markets, and data privacy.

10.4 Directions of Future Research

The future of construction lies in harnessing the potential of emerging technologies. Expanding on the identified future research directions reveals a multidimensional landscape of opportunities:

Digital Twins Beyond Construction: Future research could explore extending the application of digital twins beyond the construction phase to encompass the entire lifecycle of a building, including operation, maintenance, and eventual deconstruction. This holistic approach would allow for a more sustainable and efficient management of built assets, integrating IoT data, and AI for real-time analytics and decision-making.

AI and Machine Learning Deep Dive: Delving deeper into AI, future research might focus on the development of advanced machine learning algorithms tailored to construction-specific challenges, such as materials optimization, labour allocation, and safety monitoring. Additionally, exploring the ethical implications of AI in construction, including bias in algorithmic decision-making and job displacement, would provide a more nuanced understanding of AI's impact.

Generative AI and Collaborative Design: Investigating generative AI's potential to facilitate collaborative design processes could be a groundbreaking area of research. This would involve developing platforms where architects, engineers, and stakeholders can interact with AI-generated designs in real-time, iterating towards optimal solutions that balance aesthetic, functional, and sustainability criteria.

Integration of Digital Technologies: A critical area for future research lies in the seamless integration of various digital technologies, such as BIM, digital twins, AI, and generative design tools, within a unified construction management framework. This research could explore interoperability standards, data exchange protocols, and collaborative platforms that enable a more cohesive and efficient digital construction ecosystem.

Impact Assessment and Value Realization: Future studies should also aim to quantify the impact of digital technologies on construction productivity, cost, quality, and sustainability. Developing comprehensive metrics and models to assess the value realized from digital technology investments in construction projects would provide clearer guidance for stakeholders on prioritizing technology adoption strategies.

By addressing these expanded limitations and future research directions, the construction industry can better navigate the challenges and opportunities presented by the digital age. This forward-looking approach will not only enhance the productivity and sustainability of construction projects but also ensure that the industry remains adaptable and resilient in the face of rapid technological change.

References

AABM (2016) Aerial ABM: aerial additive building manufacturing. http://www.aerial-abm.com/

Abanda, F.H. and Byers, L. (2016) 'An investigation of the impact of building orientation on energy consumption in a domestic building using emerging BIM', *Journal of Energy*, 97, pp. 517–527.

Abanda, F.H., Mzyece, D., Oti, A.H. and Manjia, M.B. (2018) 'A study of the potential of Cloud/Mobile BIM for the management of construction projects', *Applied System Innovation*, 9(1), pp. 1–19.

Abanda, F.H., Vidalakis, C., Oti, A.H. and Tah, J.H.M. (2015) 'A critical analysis of Building Information Modelling systems used in construction projects', *Advances in Engineering Software*, 90, pp. 183–201.

Abanda, F.H., Manjia, M.B., Pettang, C., Tah, J.H.M. and Nkeng, G.E. (2014) 'Building information modelling in Cameroon: overcoming existing challenges', *International Journal of 3-D Information Modeling*, 3(4).

Abanda, H. and Tah, J.H.M. (2014) Free and open-source building information modelling for developing countries. In: The ICT Africa 2014 Conference, October 1–4, 2014 Yaoundé, Cameroon.

Ahmed, S. (2019) 'A review on using opportunities of augmented reality and virtual reality in construction project management', *Organization, Technology and Management in Construction: An International Journal*, 11, pp. 1839–1852.

Ajayi, S.O., Oyebiyi, F. and Alaka, H. (2023) 'Facilitating compliance with BIM ISO 19650 naming convention through automation', *Journal of Engineering, Design and Technology*, 21(1), pp. 108–129.

Albeaino, G. and Gheisari, M. (2021) 'Trends, benefits, and barriers of unmanned aerial systems in the construction industry: a survey study in the United States', *ITCon*, 26, pp. 84–111.

Alicke, K., Rachor, J. and Seyfert, A. (2016) *Supply Chain 4.0 – the next-generation digital supply chain.* https://www.mckinsey.com/business-functions/operations/our-insights/supply-chain-40–the-next-generation-digital-supply-chain

Al-Saeed, Y.W. and Ahmed, A. (2018) 'Evaluating design strategies for nearly zero energy buildings in the middle East and North Africa regions', *Designs*, 2(35), pp. 1–12.

Alsamarraie, M., Ghazali, F., Hatem, Z.M. and Flaih, A.Y. (2022) 'A review on the benefits, barriers of the drone employment in the construction site', *Jurnal Teknologi*, 84(2), pp. 21–31.

Alshibani, A., Aldossary, M.S., Hassanain, M.A., Hamida, H., Aldabbagh, H. and Ouis, D. (2024) 'Investigation of the driving power of the barriers affecting BIM adoption in construction management through ISM', *Results in Engineering*, 24(102987).

Andreasen, M.M. (1987) *Integrated product development.* Springer Verlag.

Arif, M. and Egbu, C. (2010) 'Making a case for offsite construction in China', *Engineering, Construction and Architectural Management*, 17(6), pp. 536–548.

Ashworth, S., Dillinger, M. and Körkemeyer, K. (2023) 'BIM guidance to optimise the operational phase: defining information requirements based on ISO 19650', *Facilities*, 41(5/6), pp. 337–356.

Atashbar, H. and Noorzai, E. (2023) 'Optimization of exterior wall cladding materials for residential buildings using the non-dominated sorting genetic algorithm II (NSGAII) based on the integration of building information modeling (BIM) and life cycle assessment (LCA) for energy consumption: a case study', *Sustainability*, 15(21), p. 15647.

Atzori, L., Iera, A. and Morabito, G. (2017) 'Understanding the internet of things: definition, potentials, and societal role of a fast-evolving paradigm', *Ad Hoc Networks*, 56, pp. 122–140.

Autodesk (2018a) *Construction disconnected: the high cost of poor data and miscommunication.* Plangrid-FMI. Available at: http://constructionblog.autodesk.com/wp-content/uploads/2023/03/Construction_Disconnected.pdf (Accessed 27 October 2023).

Autodesk (2018b) *Select clash test options.* https://knowledge.autodesk.com/support/navisworks-products/learn-explore/caas/CloudHelp/cloudhelp/2017/ENU/Navisworks-Manage/files/GUID-B13377D4-8AFA-4E95-8435-C6DC4DF26AF4-htm.html

Azhar, S. (2011) 'Building Information Modeling (BIM): trends, benefits, risks, and challenges for the AEC industry', *Leadership and Management in Engineering*, 11(3), pp. 241–252.

Bag, S., Gupta, S. and Luo, Z. (2020) 'Examining the role of Logistics 4.0 enabled dynamic capabilities on firm performance', *The International Journal of Logistics Management*, 31(3), pp. 607–628.

Bahar, Y.N., Pere, C., Landrieu, J. and Nicolle, C. (2013) 'A thermal simulation tool for building and its interoperability through the Building Information Modeling (BIM) platform', *Buildings*, 3(2), pp. 380–398.

Barbosa, F., Mischke, J. and Parsons, M. (2017) *Improving construction productivity.* https://www.mckinsey.com/capabilities/operations/our-insights/improving-construction-productivity#/

Barnes, P. (2020) *BIM for project managers: digital construction management.* ICE Publishing.

Barnes, P. and Davies, N. (2014) *BIM in principle and practice.* ICE Publishing.

Bayomi, N. and Fernandez, J.E. (2023) 'Eyes in the sky: drones applications in the built environment under climate change challenges', *Drones*, 7(10), p. 637.

Bekele, M.K. and Champion, E. (2019) Redefining mixed reality: user-reality-virtuality and virtual heritage perspectives. Intelligent & Informed, Proceedings of the 24th International Conference of the Association for Computer-Aided Architectural Design Research in Asia (CAADRIA) 2019, Volume 2. Hong Kong: Association for Computer-Aided Architectural Design Research in Asia (CAADRIA), pp. 675–684.

Berger, R. (2016) 'Digitization in the construction industry', in *Competence center civil economics.* Roland Berger GmbH.

Berlak, J., Hafner, S. and Kuppelwieser, V.G. (2020) 'Digitalization's impacts on productivity: a model-based approach and evaluation in Germany's building construction industry', *Production Planning & Control*, 32, pp. 1–11. doi:10.1080/09537287.2020.1740815.

Bikas, C., Argyrou, A., Pintzos, G., Giannoulis, C., Sipsas, K., Papakostas, N. and Chryssolouris, G. (2016) 'An automated assembly process planning system', *Procedia CIRP*, 44, pp. 222–227.

BIM Wiki (2023) *Digital twin.* [Online] https://www.designingbuildings.co.uk/wiki/Digital_twin

Bock, T. and Linner, T. (2015) 'Robot oriented design – design and management tools for the deployment of automation and robotics in construction', in *The Cambridge handbooks in construction robotics.* Cambridge University Press.

Bolaji, W.A., Olatunde, O.T., Valentine, N.N. and Fatai, O.B. (2024) 'Investigation of the use of drones by construction professionals in addressing built environment tasks', *Green Building Construction Economics*, 5(1), pp. 1–16.

Boston Consulting Group (2020) *Embracing Industry 4.0 and rediscovering growth.* Available at: https://www.bcg.com/en-us/capabilities/operations/embracing-industry-4.0-rediscoveringgrowth (Accessed 14 August 2020).

Brandín, R. and Abrishami, S. (2024) 'IoT-BIM and blockchain integration for enhanced data traceability in offsite manufacturing', *Automation in Construction*, 159(105266).

Cao, Y., Xu, C., Aziz, N.M. and Kamaruzzaman, S.N. (2023) 'BIM–GIS integrated utilization in urban disaster management: the contributions, challenges, and future directions', *Remote Sensing*, 15(5), p. 1331.

Carey, R. (2021) Explained: Industry 4.0 and the construction industry. https://www.asite.com/blogs/explained-industry-4.0-and-the-construction-industry

Carter, J. (2019) Hacked driverless cars could cause collisions and gridlock in cities, say researchers. *Forbes.* https://www.forbes.com/sites/jamiecartereurope/2019/03/05/hacked-driverless-cars-could-cause-collisions-and-gridlock-in-cities-say-researchers/?sh=4f4c8d112a09

CDBB (2018) *The Gemini principles.* Centre for Digital Built Britain.

Chang, J., Xie, X. and Parlikad, A.K. (2024) 'Cloud-based Building Information Modelling (CBIM): software engineers' insights into improving asset Information quality', *Built Environment Project and Asset Management.* https://doi.org/10.1108/BEPAM-12-2023-0247

Chen, Q., Garcia de Soto, B. and Adley, B.T. (2018) 'Construction automation: research areas, industry concerns and suggestions for advancement', *Automation in Construction*, 94, pp. 22–38.

Chen, Y., Wang, X., Liu, Z., Cui, J., Osmani, M. and Demian, P. (2023) 'Exploring Building Information Modeling (BIM) and Internet of Things (IoT) integration for sustainable building', *Buildings*, 13(2), p. 288.

Chen, Z., Han, C., Gao, M., Kandukuri, S.Y. and Zhou, K. (2021) 'A review on qualification and certification for metal additive manufacturing', *Virtual and Physical Prototyping*, 17(2), pp. 382–405. https://doi.org/10.1080/17452759.2021.2018938.

Cheng, Q., Tayeh, B.A., Abu Aisheh, Y.I., Alaloul, W.S. and Aldahdooh, Z.A. (2024) 'Leveraging BIM for sustainable construction: benefits, barriers, and best practices', *Sustainability*, 16(17), p. 7654.

Choi, H.-W., Kim, H.-J., Kim, S.-K. and Na, W.S. (2023) 'An overview of drone applications in the construction industry', *Drones*, 7(8), p. 515.

Choi, W., Na, S. and Heo, S. (2024) 'Integrating drone imagery and AI for improved construction site management through building information modeling', *Buildings*, 14(4), p. 1106.

Christoforou, E., Kylili, A., Fokaides, P.A. and Ioannou, I. (2016) 'Cradle to site Life Cycle Assessment (LCA) of adobe bricks', *The Journal of Cleaner Production*, 112(1), pp. 443–452.

Conner, B.P., Manogharan, G.P., Martof, A.N., Rodomsky, L.M., Rodomsky, C.M., Jordan, D.C. and Limperos, J.W. (2014) 'Making sense of 3-D printing: creating a map of additive manufacturing products and services', *Additive Manufacturing*, 1–4, pp. 64–76.

Construction Robotics (2019) *SAM100*. Available at: https://www.planswift.com/blog/semi-automated-robot-sam100/ (Accessed 03 September 2025)

Craveiro, F., Duarte, J.P., Bartolo, H. and Bartolo, P.J. (2019) 'Additive manufacturing as an enabling technology for digital construction: a perspective on Construction 4.0', *Automation in Construction*, 103, pp. 251–267. doi:10.1016/j.autcon.2019.03.011.

Craveiroa, F., Duartec, J.P., Bartoloa, H. and Bartolod, P.J. (2019) 'Additive manufacturing as an enabling technology for digital construction: a perspective on Construction 4.0', *Sustainable Development*, 4, p. 6. doi:10.1016/j.autcon.2019.03.011.

Cropp, C. (2020) *BIM-to-field workflows*. https://info.vercator.com/blog/bim-to-field-workflows

Dallasega, P., Rauch, E. and Linder, C. (2018) 'Industry 4.0 as an enabler of proximity for construction supply chains: a systematic literature review', *Computer Industry*, 99, pp. 205–225. doi:10.1016/j.compind.2018.03.039.

Daniotti, B., Gianinetto and Torre, S.D.(eds.) (2020) *Digital transformation of the design, construction and management processes of the built environment*. Springer Open.

Das, M., Tao, X. and Cheng, J.C.P. (2021) 'BIM security: a critical review and recommendations using encryption strategy and blockchain', *Automation in Construction*, 126(103682).

DBW (2022) *Refurbishment*. https://www.designingbuildings.co.uk/wiki/Refurbishment

DesignBuildings Wiki (2021) *ConTec*. https://www.designingbuildings.co.uk/wiki/ConTech

Ding, L., Wei, R. and Che, H. (2014) 'Development of a BIM-based automated construction system', *Procedia Engineering*, 85(2014), pp. 123–131.

Dörfler, K., Dielemans, G., Leutenegger, S., Jenny, S.E., Pankert, J., Sustarevas, J., Lachmayer, L., Raatz, A. and Lowke, D. (2024) 'Advancing construction in existing contexts: prospects and barriers of 3D printing with mobile robots for building maintenance and repair', *Cement and Concrete Research*, 186, p. 107656.

Essers, M.S. and Vaneker, T.H.J. (2013) 'Developing concepts for improved efficiency of robot work preparation', *Procedia CIRP*, 7, pp. 515–520.

Eze, E.C., Sofolahan, O., Uzoma, C.N., Ameyaw, E.E. and Omoboye, O. (2024) 'Impediments to building information modelling-enabled construction waste management in Nigeria', *Built Environment Project and Asset Management*. https://doi.org/10.1108/BEPAM-12-2023-0217

Fan, J. and Saadeghvaziri, M.A. (2019) 'Applications of drones in infrastructures: challenges and opportunities', *International Journal of Mechanical and Mechatronics Engineering*, 13(10), pp. 649–655.

Farouk, A.M., Naganathan, H., Rahman, R.A. and Kim, J. (2024) 'Exploring the economic viability of virtual reality in architectural, engineering, and construction education', *Buildings*, 14(9), p. 2655.

FIEC (2022) *Digitalisation, Construction 4.0 and BIM*. https://www.fiec.eu/priorities/digitalisation-construction-40-and-bim

Ford, S. and Despeisse, M. (2016) 'Additive manufacturing and sustainability: an exploratory study of the advantages and challenges', *Journal of Cleaner Production*, 137, pp. 1573–1587.

Forgues, D. (2019) De l'Industrie 4.0 à la Construction 4.0: Des exemples concrets! Congrès 2019: Influencez l´avenir.

Frank, A.G., Dalenogare, L.S. and Ayala, N.F. (2019) 'Industry 4.0 technologies: implementation patterns in manufacturing companies', *International Journal of Production Economics*, 210, pp. 15–26.

Furness, A. (2000) 'Machine-readable data carriers – a brief introduction to automatic identification and data capture', *Assembly Automation*, 20(1), pp. 28–34.

Fu, S., Yang, D., Mei, Z. and Zheng, W. (2025) 'Progress in construction robot path-planning algorithms: Review', *Applied Sciences*, 15(3), p. 1165. doi:10.3390/app15031165.

Gartner (2022) *Gartner glossary*. https://www.gartner.com/en/glossary

GBCA and thinkstep-anz (2021) *Embodied carbon and embodied energy in Australia's buildings*. Green Building Council of Australia and thinkstep-anz.

Gheisari, M. and Esmaeili, B. (2019) 'Applications and requirements of unmanned aerial systems (UASs) for construction safety', *Safety Science*, 118, pp. 230–240.

Ghosh, B. and Karmakar, S. (2024) '3D printing technology and future of construction: a review', *IOP Conference Series: Earth and Environmental Science*, 1326, p. 012001.

Graetz, G. and Michaels, G. (2018) 'Robots at work', *The Review of Economics and Statistics*, 100(5), pp. 753–768.

Gronkbot 3D Printing. (2019) *3D printer basics*. [online] Available at: https://gronkwena.wordpress.com/2013/04/18/3d-printer-basics/ (Accessed 19 March 2019).

Guenzi, P. and Habel, J. (2020) 'Mastering the digital transformation of sales', *California Management Review*, 62(4), pp. 57–85. doi:10.1177/0008125620931857.

Hamma-Adama, M. and Kouider, T. (2019) 'What are the barriers and drivers towards BIM adoption in Nigeria?' in Skibniewski, M.J. and Hajdu, M. (eds.) *Proceedings of the 2019 Creative Construction Conference (CCC 2019), 29 June–2 July 2019, Budapest, Hungary.* Diamond Congress.

Hauashdh, A., Nagapan, S., Jailani, J. and Gamil, Y. (2024) 'An integrated framework for sustainable and efficient building maintenance operations aligning with climate change, SDGs, and emerging technology', *Results in Engineering*, 21, p. 101822.

Heusler, W. and Kadija, K. (2018) 'Advanced design of complex façades', *Intelligent Buildings International*, 10(4), pp. 220–233. doi:10.1080/17508975.2018.1493979.

Hisham, S. (2018) Is construction industry ready for City Information Modeling? Geospatial Word. https://www.geospatialworld.net/blogs/is-construction-industry-ready-for-city-information-modeling/

Hu, X., Olgun, G. and Assaad, R.H. (2024) 'An intelligent BIM-enabled digital twin framework for real-time structural health monitoring using wireless IoT sensing, digital signal processing, and structural analysis', *Expert Systems With Applications*, 252(Part A).

Hvam, L., Mortensen, N.H. and Riis, J. (2008) *Product customization*. Springer.

Jowett, A. (2019) *The digital twin issue; Putting theory into practice*. BIMTODAY.

Kayode, O. and Olusegun, A.E. (2013) 'Local building materials: a tool towards effective low-income housing in Nigeria', *Middle-East Journal of Scientific Research*, 18(4), pp. 492–497.

Kekana, G., Aigbavboa, C. and Thwala, W. (2015) Understanding building information modelling in the South Africa construction industry. In: OTMC Conference.

Keskin, B., Salman, B. and Ozorhon, B. (2020) 'Airport project delivery within BIM-centric construction technology ecosystems', *Engineering, Construction and Architectural Management*, 28(2), pp. 530–548.

Khosrowshahi, F. (2017) 'Building information modelling (BIM) a paradigm shift in construction', in Dastbaz, M., Gorse, C., Moncaster, A. (eds.) *Building information modelling, building performance, design and smart construction.* Springer. https://doi.org/10.1007/978-3-319-50346-2_4

Kim, J., Chung, D., Kim, Y. and Kim, H. (2022) 'Deep learning-based 3D reconstruction of scaffolds using a robot dog', *Automation in Construction*, 134.

Kim, J.-Y., Lee, D. and Kim, G.-H. (2024) 'Measurement of work progress using a 3D laser scanner in a structural framework for sustainable construction management', *Sustainability*, 16(3), p. 1215.

Kim, K., Kim, H. and Kim, H. (2017) 'Image-based construction hazard avoidance system using augmented reality in wearable device', *Automation in Construction*, 83, pp. 390–403.

Korane, K. (2017) *Volvo unveils next-generation autonomous hauler*. Available at: https://www.fluidpowerworld.com/volvo-unveils-next-generation-autonomous-hauler/ (Accessed 16 February 2019).

Kousi, N., Christos Gkournelos, C., Aivaliotis, S., Giannoulis, C., Michalos, G. and Makris, S. (2019) 'Digital twin for adaptation of robots' behavior in flexible robotic assembly lines', *Procedia Manufacturing*, 28, pp. 121–126.

Kozlovska, M., Petkanic, S., Vranay, F. and Vranay, D. (2023) 'Enhancing energy efficiency and building performance through BEMS-BIM integration', *Energies*, 16(17), p. 6327.

Krause, F.L., Kimura, F. and Kjellberg, T. (1993) 'Product modelling', *Annals of the CIRP*, 42(2), pp. 695–706.

LaserDesign (2022) *3D scanning technology – hard work that looks like "magic"*. https://www.laserdesign.com/what-is-3d-scanning

Lee, J.S., Ham, Y., Park, H. and Kim, J. (2022) 'Challenges, tasks, and opportunities in teleoperation of excavator toward human-in-the-loop construction automation', *Automation in Construction*, 135 (104119).

Lee, G., Sacks, R., & Eastman, C. M. (2006) 'Specifying parametric building object behavior (BOB) for a building information modeling system', *Automation in Construction*, 15(6), 758–776.

LetsBuild (2023) *A history of BIM*. https://www.letsbuild.com/blog/a-history-of-bim

LETI (2020). *Climate emergency design guide*. London Energy Transformation Initiative. Retrieved from https://www.leti.london/cedg

Li, J., Greenwood, D. and Kassem, M. (2019) 'Blockchain in the built environment and construction industry: a systematic review, conceptual models and practical use cases', *Automation in Construction*, 102, pp. 288–307. doi:10.1016/j.autcon.2019.02.005.

Liang, P. and Sharudin, S.A. (2023) 'The application and prospects of digital visualization in urban public building space design', *World Journal of Engineering and Technology*, 11(04), pp. 976–988.

LiDAR UK (2019) *Find out all you need to know about LiDAR, its history, its uses and the technologies behind it...* Available at: http://www.lidar-uk.com (Accessed 16 February 2019).

Linner, T. and Bock, T. (2012) 'Evolution of large-scale industrialisation and service innovation in Japanese prefabrication industry', *Construction Innovation*, 12(2), pp. 156–178.

Lundeen, K., Kamat, V.R., Menassa, C.C. and McGee, W. (2019) 'Autonomous motion planning and task execution in geometrically adaptive robotized construction work', *Automation in Construction*, 100, pp. 24–45.

Mahbub, R. (2008) An Investigation into the Barriers to the Implementation of Automation and Robotics Technologies in the Construction Industry. PhD thesis, Queensland University of Technology.

Malla, V., Tummalapudi, M. and Delhi, V.S.K. (2024) 'Perceptions of built-environment professionals on using ISO 19650 standards for information management', *Journal of Legal Affairs and Dispute Resolution in Engineering and Construction*, 16(1).

Mandal, S. (2013) 'Brief introduction of virtual reality & its challenges', *International Journal of Scientific & Engineering Research*, 4(4), pp. 304–309.

Maskuriy, R., Selamat, A., Maresova, P., Krejcar, O. and Olalekan, O.O. (2019) 'Industry 4.0 for the construction industry: review of management perspective', *Economies*, 7(3), p. 68. doi:10.3390/economies7030068.

MB (2022) *What are the different types of 3D printing technologies and 3D printers?* https://www.makerbot.com/stories/design/types-of-3d-printing-technologies/

McKinsey & Company (2019) *The impact and opportunities of automation in construction*. https://www.mckinsey.com/business-functions/operations/our-insights/the-impact-and-opportunities-of-automation-in-construction

Mills, F. (2016a) *What is digital construction?* https://www.theb1m.com/video/what-is-digital-construction

Mills, F. (2016b) *Why do it in construction*. https://www.theb1m.com/video/what-is-digital-construction

Monla, Z., Beladjine, D., Assila, A. and Zghal, M. (2024) 'Indicators specification for maturity evaluation of BIM-based VR/AR systems using ISO/IEC 15939 standard', in De Paolis, L.T., Arpaia, P., Sacco, M. (eds.) *Extended reality. XR Salento 2024. Lecture notes in computer science*, vol 15027. Springer. https://doi.org/10.1007/978-3-031-71707-9_19

Nadim, W. and Goulding, J.S. (2011) 'Offsite production: a model for building down barriers: a European construction industry perspective', *Engineering, Construction and Architectural Management*, 18(1), pp. 82–101.

National Ocean Science (2021) *What is Lidar?* https://oceanservice.noaa.gov/facts/lidar.html

Nhachi, T. and Abanda, F.H. (2017) Cloud BIM collaboration tools for building and infrastructure development in Africa. In: ICEASSM, 18th - 21st April, Accra, Ghana.

Oesterreich, T.D. and Teuteberg, F. (2016) 'Understanding the implications of digitisation and automation in the context of Industry 4.0: a triangulation approach and elements of a research agenda for the construction industry', *Computers in Industry*, 83, pp. 121–139.

Olanipekun, A.O. and Sutrisna, M. (2021) 'Facilitating digital transformation in construction – a systematic review of the current state of the art', *Frontiers in Built Environment*. https://doi.org/10.3389/fbuil.2021.660758

Olanrewaju, O.I., Chileshe, N., Babarinde, S.A. and Sandanayake, M. (2020) 'Investigating the barriers to Building Information Modelling (BIM) implementation within the Nigerian construction industry', *Engineering, Construction and Architectural Management*, 27, pp. 2931–2958.

Pacheco, A., Pacheco-Pumaleque, L., Uribe-Hernández, Y., Mogrovejo, A., Pariona-Luque, R., Añaños-Bedriñana, M., Alvarez, A., Peñaranda, C., Pineda, F., Ruiz, M. and Felix-Poicon, E. (2024) 'Transforming construction management in Peru: the role of BIM in innovation and efficiency', *Sage Open*, 14(1), pp. 1–12.

Paliga, M. and Pollak, A. (2021) 'Development and validation of the fluency in human – robot interaction scale. A two-wave study on three perspectives of fluency', *International Journal of Human-Computer Studies*, 155(102698). http://doi.org/10.1016/j.ijhcs.2021.102698.

Pan, W., Gibb, A.G. and Dainty, A.R. (2012) 'Strategies for integrating the use of off-site production technologies in house building', *Engineering, Construction and Architectural Management*, 138(11), pp. 1331–1340.

Park, J., Kim, S. and Suh, K. (2018) 'A comparative analysis of the environmental benefits of drone-based delivery services in urban and rural areas', *Sustainability*, 10(888). https://doi.org/10.3390/su10030888.

PD 19650-0:2019 (2019) *Transition guidance to BS EN ISO 19650*. BSI.

Pitcher, G. (2018) *Civils 4.0 | future present. New civil engineer*. https://www.newcivilengineer.com/archive/civils-4-0-future-present-18-09-2018/

Raaj, M. (2021) *Digitalisation vs digitalization. What is the difference & how does it matter for B2B businesses?* https://www.linkedin.com/pulse/digitalisation-vs-digitalization-what-difference-how-does-melvin-raaj

Raitviir, C.-R. and Lill, I. (2024) 'Conceptual framework of information flow synchronization throughout the building lifecycle', *Buildings*, 14(7), p. 2207.

Rezaei, R., Chiew, T.K. and Lee, S.P. (2014) 'A review on E-business interoperability frameworks', *Journal of Systems and Software*, 93, pp. 199–216.

RICS (2017). *Whole life carbon assessment for the built environment* (1st ed.) Royal Institute of Chartered Surveyors, UK. https://www.rics.org/content/dam/ricsglobal/documents/standards/whole_life_carbon_assessment_for_the_built_environment_1st_edition_rics.pdf

Riverview (2024) *Riverview affordable housing strategy 2024-2029*. https://www.townofriverview.ca/sites/default/files/2024-09/AffordableHousingStrategy_2024.pdf

Rogers, H., Srivastava, M., Tsakou, M.E. and Prashad, A. (2024) 'Sustainable supply chain models for 3D printed houses', *Supply Chain Forum: An International Journal*, 1–16. https://doi.org/10.1080/16258312.2023.2298186.

Rojko, A. (2017) 'Industry 4.0 concept: background and overview', *International Journal of Interactive Mobile Technologies*, 11(5), pp. 77–90. doi:10.3991/ijim.v11i5.7072.

Rondeau, E.P., Brown, R.K. and Lapides, P.D. (2012) *Facility management*. John Wiley & Sons.

Ross, J. (2017) Don't confuse digital with digitization. *MIT Sloan Management Review*. https://sloanreview.mit.edu/article/dont-confuse-digital-with-digitization/

Sacks, R., Perlman, A. and Barak, R. (2013) 'Construction safety training using immersive virtual reality', *Construction Management Economics*, 31(9), pp. 1005–1017.

Sacks, R., Wang, Z., Ouyang, B., Utkucu, D. and Chen, S. (2022) 'Toward artificially intelligent cloud-based building information modelling for collaborative multidisciplinary design', *Advanced Engineering Informatics*, 53, p. 101711.

Saka, A.B. and Chan, D.W.M. (2019) 'A scientometric review and metasynthesis of building information modelling (BIM) research in Africa', *Buildings*, 9(85), pp. 1–21.

Salem, T., Dragomir, M. and Chatelet, E. (2024) 'Strategic integration of drone technology and digital twins for optimal construction project management', *Applied Sciences*, 14(11), p. 4787.

Salesforce (2020) *What is digital transformation?* Available at: https://www.salesforce.com/products/platform/what-is-digital-transformation/ (Accessed 12 December 2020). [Google Scholar]

Schmarzo, B. (2017) *What is digital transformation?* https://www.linkedin.com/pulse/what-digital-transformation-bill-schmarzo/

Schober, K.-S. (2016) *Digitization in the construction industry: building Europe's road to "Construction 4.0"*. Roland Berger GMBH.

Schwab, K. (2015) *The fourth industrial revolution: what it means and how to respond*. Foreign Affairs. https://www.foreignaffairs.com/articles/2015-12-12/fourth-industrial-revolution

Shan, J. and Toth, C.K. (2018) *Topographic laser ranging and scanning: principles and processing*. 2nd edn. CRC Press.

Sharma, L.K. and Naik, R. (2024) 'Present technologies for wetland studies', in *Conservation of saline wetland ecosystems*. Springer.

Shaw, J. (2019) 'Platform real estate: theory and practice of new real estate markets', *Urban Geography*, 41(8). doi:10.1080/02723638.2018.1524653.

Shea, S. and Burns, E. (2020) *Smart city*. https://www.techtarget.com/iotagenda/definition/smart-city#:~:text=A%20smart%20city%20is%20a,government%20services%20and%20citizen%20welfare.

Shrivastava, S. and Chini, A. (2012) 'Using building information modelling to assess the initial embodied energy of a building', *International Journal of Construction Management*, 12(1), pp. 51–64.

Skibniewski (1992) Current status of construction automation and robotics in the United States of America. The 9th International Symposium on Automation and Robotics in Construction (1992), pp. 17–26.

Smith, P. (2014) 'BIM implementation-global strategies', *Procedia Engineering*, 85, pp. 482–492.

Sovetova, M. and Calautit, J.K. (2024) 'Thermal and energy efficiency in 3D-printed buildings: review of geometric design, materials and printing processes', *Energy and Buildings*, 323, p. 114731.

Starr, C.W., Saginor, J. and Worzala, E. (2021) 'The rise of PropTech: emerging industrial technologies and their impact on real estate', *Journal of Property Investment & Finance*, 39(2), pp. 157–169.

Szokolay, S.V. (2008) *Introduction to architectural science: the basis of sustainable design*. Elsevier Ltd.

Taylor, T.S. (2019) *Introduction to laser science and engineering*. CRC Press.

Teicholz, P. (2013) *BIM for facility managers*. John Wiley & Sons.

The Construction Index (2022) *Lendlease team gets grant to develop 'Building 4.0'*. https://www.theconstructionindex.co.uk/news/view/lendlease-team-gets-grant-to-develop-building-40

The House (2018) *UK needs to retrofit 26 million homes by 2050 to reduce greenhouse gas emissions*. https://citizen-network.org/library/retrofitting-listed-buildings.html#:~:text=There%20are%20approximately%2026%20million,retrofitted%20at%20the%20same%20time%2C

Thurgood, J.D. and Bethel, J.S. (2003) 'Geographic information system', in Chen, W.F. and Richard Liew, J.Y. (eds.) *The civil engineering handbook*. CRC Press

Tushaj, E. (2021). *BIM adoption for structural engineering design* (Master's dissertation, Politecnico di Milano–BIM A+, Erasmus Mundus Joint Master in BIM A+). Retrieved from https://bimaplus.org/wp-content/uploads/2021/10/2021-ErsilioTushaj-Dissertation.pdf

TWI (2022) *Industry 4.0*. https://www.twi-global.com/what-we-do/research-and-technology/technologies/industry-4-0

UK Connect (2021) *10 construction technology trends impacting the industry in 2022*. https://ukconnect.com/construction-technology-trends/

UN-Habitat (2010) 'The state of African cities 2010 governance, inequality and urban land markets', in *United Nations Human Settlements Programme*. UN-HABITAT.

UN-Habitat (2016) *Supporting national and city-wide slum upgrading and prevention through national urban policy*. https://icnup.urbanpolicyplatform.org/wp-content/uploads/2016/12/08_A4_Thematic-Quick-Guide_NUPPSUP-1.pdf

United BIM (2022) *Leading countries with BIM adoption*. https://www.united-bim.com/leading-countries-with-bim-adoption/

United Nations (2018) *68% of the world population projected to live in urban areas by 2050*. https://www.un.org/development/desa/en/news/population/2018-revision-of-world-urbanization-prospects.html

Valero, E. and Adán, A. (2016) 'Integration of RFID with other technologies in construction', *Measurement*, 94, pp. 614–620. https://doi.org/10.1016/j.measurement.2016.08.037

Vernon, J. and Hughes, J. (2017) 'Using automation to combat the impending labor shortage'. *Industry Week*. https://www.industryweek.com/supply-chain/supply-chain-technology/article/22024093/using-automation-to-combat-the-impending-labor-shortage

Vishweshwaran, M. and Sujatha, E.R. (2024) 'A review on applications of drones in geotechnical engineering', *Indian Geotechnical Journal*, 55, pp. 2091–2105.

Vivian, M. (2022) *How BIM and VR in construction are boosting the industry*. https://vection-technologies.com/blog/how-bim-and-vr-in-construction-are-boosting-the-industry/

Wang, C., Adetola, S.H. and Abdul-Rahman, H. (2015) 'Assessment of BIM implementation among MEP firms in Nigeria', *International Journal of Advances in Applied Sciences*, 4(3), pp. 73–81.

Wang, X. et al. (2018) 'A new grid map construction method for autonomous vehicles', *IFAC-PapersOnLine*, 51(31), pp. 377–382.

Wastiels, L. and Decuypere, R. (2019). 'Identification and comparison of LCA–BIM integration strategies', *IOP Conference Series: Earth and Environmental Science*, 323(1), p. 012101. doi:10.1088/1755-1315/323/1/012101

Winkelhaus, S. and Grosse, E.H. (2020) 'Logistics 4.0: a systematic review towards a new logistics system', *International Journal of Production Research*, 58(1), pp. 18–43.

Wray, S. (2021) *Pandemic gives rise to Cities 4.0, where smart meets sustainable*. https://cities-today.com/pandemic-gives-rise-to-cities-4-0-where-smart-meets-sustainable/

Yönder, V.M. and Çavka, H.B. (2024) 'A case study of clash detection for early design phases IN building information modelling', *Livenarch+ Journal*, 1(1), pp. 57–71.

Yunus, R. and Yang, J. (2012) 'Critical sustainability factors of industrialised building systems', *Construction Innovation*, 12(4), pp. 447–463.

Zhai, X., Reed, R. and Mills, A. (2014) 'Factors impeding the offsite production of housing construction in China: an investigation of current practice', *Construction Management and Economics*, 32(1–2), pp. 40–52.

Zhou, S. and Gheisari, M. (2018) 'Unmanned aerial system applications in construction: a systematic review', *Construction Innovation*, CI-02-2018-0010.

Zhu, A., Pauwels, P., Torta, E., Zhang, H. and De Vries, B. (2024) 'Data linking and interaction between BIM and robotic operating system (ROS) for flexible construction planning', *Automation in Construction*, 163.

Index

For Product Safety Concerns and Information please contact our EU representative GPSR@taylorandfrancis.com Taylor & Francis Verlag GmbH, Kaufingerstraße 24, 80331 München, Germany